SpringerBriefs in Space D

Series Editor

Joseph N. Pelton

For further volumes:
http://www.springer.com/series/10058

Fabio Tronchetti

Fundamentals of Space Law and Policy

Fabio Tronchetti
School of Law
Harbin Institute of Technology
Harbin
People's Republic of China

ISSN 2191-8171 ISSN 2191-818X (electronic)
ISBN 978-1-4614-7869-0 ISBN 978-1-4614-7870-6 (eBook)
DOI 10.1007/978-1-4614-7870-6
Springer New York Heidelberg Dordrecht London

Library of Congress Control Number: 2013939566

Springer is part of Springer Science+Business Media (www.springer.com)

This Springer book is published in collaboration with the International Space University. At its central campus in Strasbourg, France, and at various locations around the world, the ISU provides graduate-level training to the future leaders of the global space community. The university offers a 2-month Space Studies Program, a 5-week Southern Hemisphere Program, a 1-year Executive MBA and a 1-year Masters program related to space science, space engineering, systems engineering, space policy and law, business and management, and space and society.

These programs give international graduate students and young space professionals the opportunity to learn while solving complex problems in an intercultural environment. Since its founding in 1987, the International Space University has graduated more than 3,000 students from 100 countries, creating an international network of professionals and leaders. ISU faculty and lecturers from around the world have published hundreds of books and articles on space exploration, applications, science and development.

Preface

Overview

This book is a concise introductory overview of international space law and policy. It seeks to address an audience relatively new to these fields. The objective of this short book is to cover in simple language the fundamentals of space law and policy and address key pending issues that are relevant to space law and policy experts.

This book provides the legal and political foundations of space activities as well as offering insight on present and future space law and policy trends, challenges, and opportunities. It serves as an excellent tool for those working with civil, commercial, and military space personnel and for anybody interested in these fields. A famed physicist one said that if you cannot explain a concept to someone new to your field, you do not understand it yourself. This book tried to take this admonition to heart by being as clear as possible.

The book is divided into two main parts. The first part deals with Space Law, and the second deals with Space Policy. The former describes the national and international legal frameworks governing space activities and the subjects involved in its formulation and implementation. The latter analyzes the political dimension of space activities and their impact on social, economic, and security matters. The conclusions of the book recount the main points and the way forward by recommending further reading on the subject.

Read in conjunction with the other books in the Springer Space Development series, one can indeed build a broader understanding of the business, economics, law, and policies of space activities.

What is Space Law?

Law is defined as "any system of regulations to govern the conduct of the people of a community, society or nation, in response to the need for regularity, consistency, and justice based upon collective human experience".[1] In particular, laws are made to achieve desired goals. In democratic institutions in the twenty-first century these objectives include peace, social cohesion and societal advancement, the balancing of diverging interests, and the avoidance of undesired and dangerous conditions.

One might think that this means that law—and in this case space law—is thus a boring and arcane subject. This is simply not the case. Space law addresses advanced, state-of-the art technology that is constantly evolving in new directions. It also involves the views and sometimes conflicting opinions from nations around the world about how to conduct space-related affairs. It seeks to develop processes that can be used to settle disputes. Space law is also about how to explore, utilize, and protect outer space, not only for today but for generations yet to come. Finding new solutions to complex problems in a global context is what space law is fundamentally all about.

Thus in many ways space law is exciting, stimulating, path- and precedent-setting, and sometimes quite rewarding—especially when new solutions are found to difficult issues. Although the intricacies of space law treaties and conventions might be a bit heavy going at times, this short book seeks to hit the highlights. The goal is not to be a definitive book on space law but rather to explain the major features of space law and associated space policy. Some of the more important issues currently pending in the early part of the twenty-first century in the field of space law will be explained; for example, the pathway that starts with the so-called "soft law," including accepted practices and codes of conduct, that over time can evolve into formal agreements among nations having force of law will be discussed.

Broadly speaking the term "space law" is used with reference to the set of international and national rules and regulations governing human activities in and relating to outer space.[2] The purpose of space law is to establish a legal environment enabling the achievement of common goals and interests related to the exploration and use of outer space; at the same time, it aims at preventing the emergence of tensions and conflicts among the subjects involved in outer space activities.[3]

As a starting point, we can identify three main facets of space law. These are: its scope, its fragmentation, and its evolutionary nature.

1. Scope: Space law is applicable not only to activities taking in place in outer space, for example the collection of images and data by a satellite, but also to events occurring on Earth that are related to outer space, i.e., liability for damage caused by a space object or a part of it falling to the ground.

Even if it might appear surprising, international space law does not include a definition of "outer space," nor gives a precise indication of where outer space begins. Scholars and diplomats have been unable to reach an agreement on these two points since the beginning of the Space Age. Nevertheless, many argue[4] that the lower border of outer space should be set at an altitude of 100 km above sea level (62.5 miles).[5]

2. Fragmentation: Although there is a central body of laws, namely the five U. N. space treaties, space law does not exist as a single, coherent, and comprehensive body of legal principles and rules governing human activities in outer space. Rather it can be seen as a 'box' containing many different types of norms to deal with the practical problems connected with the exploration and use of outer space. Consequently, regulation of space activities is achieved through amalgamation and application of all possible rules.
3. Evolutionary Nature: The body of space law has been constantly growing since the entry into force of the first international treaty on outer space, namely the 1967 Outer Space Treaty. This is the consequence of the fact that, in the past 40 years, new developments and technologies have changed the nature and dimension of space activities. In order to ensure that these activities were carried out in an orderly and peaceful manner, space law had to adapt itself to these changes and progressively evolve.

What is Space Policy?

In its ordinary interpretation the word *policy* means "a plan or course of action, as of a government, political party, or business, intended to influence and determine decisions, actions, and other matters".[6] In the context of outer space, the term refers to the official approach of a state towards the exploration and use of outer space. Normally, a "space policy" describes a nation's strategy regarding its civilian space program and the military and commercial utilization of outer space. Furthermore, space policies include both the making of space policy through the legislative process and the execution of that policy by civilian, military bodies, and regulatory agencies.

As the military, economic, and social implications of the uses of space expand, so it does the relevance of outer space on a worldwide scale. The utilization of outer space has become a global phenomenon affecting the lives of millions of people and influencing international relations. Consequently, questions related to the access and use of outer space have been placed at the core of the strategic agenda of the technologically advanced nations.

In a similar scenario national space policies acquire a special importance. On one side, they give direction to all national subjects involved in space activities. On the other side, they constitute a tool to enhance transparency over the space activities of a certain country. In this way, they also strengthen trust among space participants and, ultimately, favor international cooperation. Thus, nowadays outer space-related issues significantly influence economic, political, and military decision-making at the national and international level.

Acknowledgments

I am deeply indebted to the series editor, Dr. Joseph N. Pelton, for his support, trust and comments which significantly improved the quality of the final manuscript. His enthusiasm and quick feedbacks have largely contributed to speed up the completion of this book. I would also like to thank the International Space University (ISU), which suggested and made possible the writing of the present book, and the publisher, Springer Science+Business Media, for accepting it as part of the series "Springer Briefs in Space Development."

My appreciation also goes to my friend and colleague Dr. Michael Mineiro, who put me in contact with the series editor.

My gratitude goes to my mentor, Prof. Frans von der Dunk, University of Nebraska-Lincoln, who, back in the days of my Ph.D. studies, shared his knowledge with me and gave me the legal and methodological foundations to undertake an academic career. His teachings always constitute a source of inspiration and guidance.

A special thank goes to my family, for its continuous support, and to my wife, for her love, patience, and for taking care of me during the writing of this book. This book is dedicated to them.

Obviously, this book remains my own product and I alone bear full responsibility for the views expressed and for any errors or omissions it may contain.

April 2013

Fabio Tronchetti
Harbin, China

Contents

Part I Space Law

Part II Space Policy

Abbreviations

ABM	Anti-Ballistic Missile
APIC	Asia-Pacific International Space Year Conference
APSCO	Asia Pacific Space Cooperation Organization
APSRAF	Asia-Pacific Regional Space Agency Forum
ASAT	Anti-Satellite Test
CCIC	Commercial Crew Integrated Capability
CGWC	China Great Wall Corporation
CNES	Centre National d'Ėtudes Spatialies
DBS	Direct Broadcasting by Satellite
DOD	(U. S.) Department of Defense
EADS	European Aeronautic Defense and Space Company N.V.
ESA	European Space Agency
EU	European Union
EUMETSAT	European Organization for the Exploitation of Meteorological Satellites
EUTELSAT	European Telecommunication Satellite Organization
FAA	(U. S.) Federal Aviation Administration
FAO	Food and Agriculture Organization
FCC	(U. S.) Federal Communications Commission
GALILEO	European Global Navigation Satellite System
GMES	Global Monitoring for Environment and Security
GNSS	Global Navigation Satellite System
GPS	Global Positioning System
GSO	Geostationary Orbit
IAA	International Academy of Astronautics
IAASS	International Association for the Advancement of Space Safety
IADC	Inter-Agency Space Debris Coordination Committee
IAF	International Astronautical Federation
ICAO	International Civil Aviation Organization
ICJ	International Court of Justice
IISL	International Institute of Space Law
ILA	International Law Association

INMARSAT	International Maritime Satellite Organization
INTELSAT	International Telecommunication Satellite Consortium
ISRO	Indian Space Agency
ISS	International Space Station
ISSF	International Space Safety Foundation
ISU	International Space University
ITAR	International Traffic in Arms Regulations
ITSO	International Telecommunication Satellite Organization
ITU	International Telecommunication Union
JAXA	Japanese Aerospace Exploration Agency
LEO	Low Earth Orbit
MPCV	Multi-Purpose Crew Vehicle
NASA	(U. S.) National Aeronautics and Space Administration
NPS	Nuclear Power Source
PCA	Permanent Court of Arbitration
RLV	Reusable Launch Vehicle
STS	Space Transportation System
UK	United Kingdom
UN	United Nations
UNCITRAL	United Nations Commission on International Trade Law
UNCOPUOS	United Nations Committee on the Peaceful Uses of Outer Space
UNEP	United Nations Environment Program
UNESCO	United Nations Educational, Scientific and Cultural Organization
UNGA	United Nations General Assembly
UNIDIR	United Nations Institute for Disarmament Research
UNIDROIT	International Institute for the Unification of Private Law
UNOOSA	United Nations Office for Outer Space Affairs
US	United States (of America)
USML	United States Munitions List
WARC	World Administrative Radio Conference
WMD	Weapons of Mass Destruction
WMO	World Meteorological Organization
WTO	World Trade Organization

Part I
Space Law

Chapter 1
The Legal Framework Regulating International Outer Space Activities

Sources of Outer Space Law

The analysis of any legal framework must start with the description of its sources. A source of law is where one looks to determine the law on a particular matter.[7] Within a national legal system the basic sources are the Constitution of that particular state and the legislation adopted pursuant to it.

Space law offers a rather composite picture; indeed, it includes a variety of sources to be found both at international and national levels.

The core of the whole space legal system is constituted by the instruments negotiated within the framework of the United Nations (U. N.), particularly within the U. N. Committee on the Peaceful Uses of Outer Space (UNCOPUOS). These instruments are: (a) the five U. N. space treaties; (b) a set of U. N. General Assembly principles.

In addition to the U. N. space instruments, other norms have been concluded by states on a bilateral and multilateral basis outside of the UNCOPUOS framework. The relevant parts of the statutes of international inter-governmental organizations, such as the International Telecommunication Union (ITU) and the European Space Agency (ESA), can be mentioned as examples of this kind.

Furthermore, a growing number of countries have adopted national laws and/or internal regulations governing national space activities. National laws are meant to implement international space law norms and to make sure that private space activities do not undermine international obligations and security interests.

Finally, it should be kept in mind that international outer space law is a branch of public international law. Consequently, the fundamental rules of international law, particularly those included in the U. N. Charter, are applicable to activities in outer space.

In conclusion, space law offers a rather variegate picture.[8] Despite including a vast number of applicable rules, it cannot claim to be a comprehensive and integral legal system. In particular, its main limit consists of not addressing all issues that would be desirable for completeness. Its further advancement relies on the willingness of states to cooperate towards this goal. However, as the number of

F. Tronchetti, *Fundamentals of Space Law and Policy*,
SpringerBriefs in Space Development, DOI: 10.1007/978-1-4614-7870-6_1,

countries capable of accessing and using outer space increases, the challenges to the formation of effective space law parallely augment.

The Evolution of International Space Law

Compared to other fields of law, space law is a very young area of regulation. Many could imagine that the birth date of space law coincided with the launch of the first satellite in orbit in 1957. However, discussions and legal studies on the possibility of establishing rules regulating human activities in outer space emerged much before that time.

As early as 1903 the Russian space pioneer Tsiolkovsky published a paper in which he foresaw human expansion in outer space by using liquid fuel rockets. In 1932 the first monograph that addressed the study of space law was published by Mandl.[9] During World War II and in the years that followed significant developments in rocket technology were made. This advance in rocketry and missiles created the expectations that manmade objects could be soon successfully launched into outer space. These expectations materialized when the Soviet *Sputnik I* became the first artificial satellite to orbit the Earth.[10] This achievement was followed by a series of further successful missions. In April 1961, Yuri Gagarin completed the first manned spaceflight, and in 1969 Neil Armstrong was the first human being to set foot on another celestial body, the Moon.

By then it had already became clear that legal rules regulating space activities were needed in order to prevent confusion and to avoid the development of practices dictated exclusively by national and strategic interests. It was feared that such practices could eventually result in conflicts among nations. However, the international community was aware of these risks. Thus, rather wisely, states decided to cooperate both in the establishment of the rules governing space activities and in their actual implementation (Fig. 1.1).

With these premises the United Nations was identified as the natural forum for the negotiation of international space law. In 1958 an *ad hoc* committee was installed by the U. N. General Assembly to deal with the problems arising from space activities and to provide space activities with an adequate legal framework.[11] On December 12,1959, this committee became a permanent body and was given the name of United Nations Committee on the Peaceful Uses of Outer Space (UNCOPUOS).[12] In 1962 the committee created two sub-committees, one dealing with legal matters, the Legal Sub-Committee, and one addressing technical issues, the Technical Sub-Committee.

Since its establishment UNCOPUOS has constituted the main body for the discussion and elaboration of the legal framework regulating space activities.[13] For example, the five space treaties and all of the U. N. space resolutions have been negotiated within it. In the last decades, however, the space law-making activity within UNCOPUOS has reached a near standstill. This has been the consequence of several factors, such as the length of UNCOPUOS's decision-making

Fig. 1.1 Yuri Gagarin, the first human to journey into outer space, on April 12, 1961 (Courtesy of ESA at http://spaceinimages.esa.int/Images/2011/03/Yuri_Gagarin5)

mechanism and the unwillingness of some states to accept new legally binding obligations in the field of space law.

Nevertheless, this has not stopped the formulation of new space-related rules. Technological progress, coupled with an increase in the number of space activities as well as of space participants, have created new problems, the management of which required new legal solutions. Consequently, new norms addressing specific issues, such as the reduction of orbital space debris, are being developed; this process has taken place as an alternative to UNCOPUOS, such as in the context of non-governmental organizations or on a bilateral or regional basis.

The next sections will be divided as follows. First, the space law-making activity within UNCOPUOS will be described. Then, the drafting of new space law rules outside of the UNCOPUOS framework will be analyzed.

UNCOPUOS Activities

The development of international space law within the United Nations can be divided into four stages. Each stage provides a specific contribution to the formulation of the legal framework governing activities in outer space.

Stage 1

The first stage, which ranged from the late 1950s until the mid 1960s, can be labeled as the 'preparatory' stage. During this period, the foundations of the legal framework managing activities in outer space were laid down. Significantly, the approach chosen by states for the initial development of space law was to first establish a corpus of general non-binding principles, and then to incorporate them into a binding treaty.

Pursuant to this approach, the General Assembly of the United Nations adopted in 1961[14] and 1963[15] two resolutions on outer space matters. These resolutions established some fundamental principles related to the human presence in space, such as the freedom to explore and use outer space, the prohibition to appropriate outer space or any of its parts, the applicability of international law, including the U. N. Charter, to space activities, etc. The principles included in these resolutions were to constitute the foundation of the first international treaty dealing with outer space issues, namely the 1967 Outer Space Treaty.

Stage 2

The second stage, which went from the late 1960s until the early 1980s, can be called the 'law-making era.' Within this timeframe the five U. N. space treaties were negotiated and entered into force.

Towards the end of the 1960s the time seemed mature for entering into legally binding instruments, i.e., multilateral treaties aimed at clarifying and progressively developing the rules applicable to outer space activities. Beginning in 1967, five international treaties were drafted. These treaties pursued different goals. While the 1967 Outer Space Treaty established general principles addressing the most significant issues related to space activities, the four remaining treaties focused on a specific topic that had previously been regulated under the Outer Space Treaty, only from a general perspective. Thus, the 1968 Rescue Agreement dealt with the status of astronauts, the 1972 Liability Convention covered the liability for damage caused by space objects, the 1975 Registration Convention addressed the issue of the 'registration of space objects' and, finally, the 1979 Moon Agreement focused on legal issues concerning the Moon and other celestial bodies.

Stage 3

The third stage, which ended in the middle of the 1990s, was the so-called soft-law stage. The term 'soft law' is used with reference to a variety of documents, such as resolutions, declarations, guidelines, codes of conduct,[16] characterized by their non-binding character. This is the key difference that distinguishes 'soft law' from 'hard law' instruments, i.e., treaties or conventions that are legally binding on states. A resolution adopted by the U. N. General Assembly is a typical example of a soft law instrument.

The third stage was characterized by the adoption by the U. N. General Assembly of four sets of non-binding principles regulating some special categories of space activities, such as the use of artificial satellites for international direct television broadcasting, remote sensing, the use of nuclear power sources in outer space, and the exploration and use of outer space for the benefit of all countries.[17] These principles were the outcome of years of negotiations within UNCOPUOS.

Importantly, at the beginning of the 1980s, the climate of international cooperation that had characterized the second stage had diminished. Furthermore, the five space treaties had exhausted the basic issues on which states were willing to undertake international legal obligations. Consequently, an alternative form to legally regulate pressing problems relating to the uses of outer space had to be found. The adoption of declarations of principles by the General Assembly was chosen as the optimal solution to further develop space law.

Stage 4

The fourth and final stage goes from the end of the 1990s until the present time. It is characterized by the assessment of the existing legal regime and by the formulation of non-binding documents based upon the rights and obligations provided for in the space treaties. In particular, the UNCOPUOS Legal Sub-Committee is undertaking efforts to broaden the acceptance of the U. N. space treaties and to evaluate their implementation. Furthermore, it has drafted two new resolutions on principles dealing with two pressing issues of the space agenda, namely the identification of the "launching state" and the registration of space objects. These resolutions have been adopted by the General Assembly of the United Nations.

Legal Instruments Negotiated Within the UNCOPUOS Framework

Space law includes five international treaties, respectively the 1967 Outer Space Treaty, the 1968 Rescue and Return Agreement, the 1972 Liability Convention, the 1975 Registration Convention and the 1979 Moon Agreement.[18] These treaties were negotiated within the Legal Sub-Committee of UNCOPUOS from the late 1960s until the end of the 1970s.

The space treaties can be differentiated in terms of the number of states parties to them and their legal form. Firstly, while the first four treaties have been ratified by a significant number of states, ranging from 50 to 100, only 13 states are parties to the Moon Agreement. Secondly, while the Outer Space Treaty is a treaty on principles containing only rules of a general nature, the remaining four treaties include more detailed provisions.

The 1967 Outer Space Treaty

The Outer Space Treaty is the cardinal instrument regulating activities in outer space.[19] It provides guidance and direction to human operations in the space environment and constitutes the basis for all legal documents, negotiated at both the international and national level, addressing outer space issues. All major space powers, including the United States, Russia, China, the United Kingdom, Germany, France, etc., are Parties to it. Although a detailed analysis of its provisions goes beyond the purpose of this book, the following paragraphs will outline the most significant principles of the treaty.

Article I of the Outer Space Treaty attributes to states-Parties the right to freely explore and use outer space, and the freedom to carry out scientific investigation. Significantly, Article I points out that the exploration and use of outer space, including the Moon and other celestial bodies, shall be carried out "for the benefit and in the interest of all countries, irrespective of their degree of economic and scientific development, and shall be the province of all mankind." This provision provides the philosophy that is at the core of space law, according to which all countries, in one way or another, shall benefit from the exploration and use of outer space, regardless of their degree of development. In practice this idea has, for example, found application in the context of disaster management activities, where space-faring states use their space technologies to help the affected countries that do not possess their own space technology.

The terms of Article I must be read in connection with those of Article III, establishing that space activities shall be carried out in accordance with international law and, in particular, the U. N. Charter, so as to maintain international peace and security and promote international cooperation and understanding. Article III is thus significant because it clarifies that, although space activities take place outside of Earth's boundaries, they shall not violate fundamental principles of international law.

Article II incorporates a basic principle of space law, namely that outer space, including the Moon and other celestial bodies, is not subject to national appropriation by claim of sovereignty, by means of use or occupation or by any other means. States are consequently forbidden from appropriating outer space and any of its parts. Despite some occasional opposing views,[20] such a prohibition also extends to non-state entities, i.e., private operators.

Article IV is the only provision of the treaty specifically addressing the issue of military activities in outer space. However, it only establishes a partial demilitarization of outer space, as Parties are specifically forbidden to place in orbit around Earth any object carrying nuclear weapons or any other weapons of mass destruction, install such weapons on celestial bodies, or station such weapons in outer space in any other manner. The placement in outer space of other kinds of weapons, such as anti-satellite weapons, or the transit of anti-ballistic missiles and rockets through space, is not forbidden. Full demilitarization, both in terms of banning the placement of weapons and the conduct of military activities, applies only to the Moon and other celestial bodies.

Article V includes principles regulating assistance to be given to astronauts in the event of accidents, distress, or emergency landing on the territory of other states or on the high seas. The terms of Article V have been further developed in the 1968 Rescue and Return Agreement.

Article VI establishes the principle of international responsibility of states for national space activities, whether such activities are carried out by governmental agencies or by non-governmental entities. Accordingly, states must authorize and continuously supervise the space activities of their non-governmental entities. In order to comply with the duties laid down in Article VI and to properly regulate the participation of public and private entities to space operations, several countries have adopted national space legislation.

Articles VII and VIII of the Outer Space Treaty contain rules on liability for damage and jurisdiction and control over space objects. These rules have provided the basis for the 1972 Liability Convention and 1975 Registration Convention and will be addressed under those headings. Here it suffices to say that while Article VII establishes that the "launching State" is internationally liable for damage caused by its space object to another Party or to its natural or juridical persons, Article VIII provides that states retain jurisdiction and control over the space objects that they have registered.

Article IX is the only article of the treaty addressing, although rather vaguely, environmental issues. Accordingly, Parties shall avoid harmful contamination of the outer space environment as well as adverse changes to the environment of Earth that might be caused by the introduction of extra-terrestrial material. Furthermore, it establishes that Parties must undertake international consultations prior to space activities they deem to be of a hazardous nature.

The remaining articles of the treaty contain clauses on international cooperation, sharing of information and entry into force, amendment of and withdrawal from the treaty.

The 1968 Rescue and Return Agreement

The 1968 Rescue and Return Agreement is an elaboration of Article V of the Outer Space Treaty.[21] The purpose of this agreement is to create the obligation for

Parties to assist and help astronauts experiencing situations of danger or distress and to elucidate the conditions under which such help should be given. The provisions of the agreement are pretty much Earth-oriented, in the sense that they deal with assistance to astronauts for Earth-related events, for example in case of emergency or unexpected landing in the territory of a state Party. They basically do not address the issue of assistance to astronauts in space; this is mostly due to the technical barriers impeding any form of significant help in space.

In recent years much attention has been paid to the legal status of non-professional participants to spaceflights, such as tourists on board orbital and sub-orbital flights.[22] The question is whether such paying passengers are entitled to the same benefits and protection attributed to professional astronauts (who normally work for their own governments). Essentially, the legal doctrine is split into two sections; on one side, there are those who refuse to equate private passengers to full fledge astronauts[23]; on the other side, there are others arguing that all on board a spaceflight should be treated equally, also based on reason of humanity.[24]

The 1972 Liability Convention

The 1972 Liability Convention elaborates upon Article VII of the Outer Space Treaty.[25] As previously described, Article VII merely set forth the principle of international liability of the launching state for damage caused by its space object. However, it leaves some key issues un-addressed, for example: (a) What is the "damage caused by a space object"? (b) Where did this "damage" occur? (c) What type of liability regime is applicable? (d) What is the procedure to be followed in case damage occurs? Considering the hazardous nature of space activities and the concrete possibility that space objects may cause damage to other space objects or on the surface of Earth, the above issues were in need of specific regulation. Consequently, Parties drafted a convention addressing international liability for damage caused by space objects.

The Convention begins by providing definitions of key terms, particularly "damage"[26] and "launching state."[27] These definitions are essential to delineate the context in which the Convention operates.

Then, the Convention addresses the issue of grounds for liability. In this respect, it makes a clear distinction between two situations—first, damage caused on the surface of Earth or to an aircraft in flight, and second, damage caused elsewhere than on the surface of Earth. With regard to the former, the Convention adopts a regime of absolute liability, while, as far as the latter is concerned, fault liability applies.

Under the regime of absolute liability states are always liable for damage caused by their space objects. Therefore, victims are not required to prove that damage is the consequence of the fault of the launching state. This provision is meant to protect subjects not involved in space activities, who suffered damage from a space object. Significantly, the launching country shall not be liable under the Convention for damage caused on the ground to its natural or juridical persons

(Article VI). If similar damage occurs the situation will be dealt internally and compensation cases be brought to national courts. Ironically, the 'absolute liability' provision has proved to be a stumbling block to solving the problem of how to undertake active removal of orbital space debris. This is because the current provisions do not allow a country to transfer liability to an entity undertaking the removal process.

Instead, in case of a collision between space objects, Parties are in a position of equality. Both have accepted the risks that space activities incorporate and, thus, liability cannot be absolute but must be based on the fault of one of the Parties involved. Clearly, under this regime a Party must be able to prove the fault of the other Party to the accident.

In case of a joint launch of a space object by two or more states, the Liability Convention establishes that they will be jointly and severally liable (Article V, para 1).

The Liability Convention is based on the idea that the launching state shall pay compensation for the damage caused by its space objects. Such a compensation shall be determined in accordance with international law and the principles of justice and equity (Article XII). Claims for compensation can only be presented at nation level, even when the subject suffering the damage is a private person. In such a case, the state will act on behalf of its nationals.

The procedure set forth in the Convention foresees that a state shall first present a claim to the launching state through diplomatic channels. This submission shall occur not later than one year following the date of the occurrence of the damage or the identification of the launching state, which is liable (Article X, para 1). If no settlement through diplomatic channels is possible the claim shall be submitted to a Claims Commission, established by the Parties concerned at the request of either Party (Article XIV). The significance of this commission is largely undermined by the fact that its decisions are only binding if the parties to the dispute have so agreed.[28]

Interestingly, the Liability Convention enables inter-governmental organization, subject to a few conditions, to enjoy effectively the same substantial rights and obligations under the Convention as individual state Parties (Article XXII). Currently, only three international inter-governmental organizations, namely the European Space Agency (ESA), the European Telecommunication Satellite Organization (EUTELSAT) and the European Organization for Exploitation of Meteorological Satellites (EUMETSAT), have made a declaration accepting the rights and duties provided for in the Convention.

The 1975 Registration Convention

Similar to the Rescue Agreement and the Liability Convention, the Registration Convention has its roots in the 1967 Outer Space Treaty. Indeed, it expands the scope and practical effects of Article VIII of that treaty.[29] As of January 2012, 57 states are Parties to it and two inter-governmental organizations, namely ESA and

EUMETSAT, have made declarations of acceptance of the rights and obligations provided for in the Convention.

One might wonder why space objects should be registered. The reason is simple: it contributes to maintain order in outer space and to preserve its peaceful nature. The advantages of a system of registration of space objects appear evident if one takes into account that: (1) It is virtually impossible to identify a spacecraft that has caused damage without a system of registration; (2) A balanced and informative system of registration helps minimizing the likelihood and suspicion of weapons being placed in orbit.

The bottom line of the Convention is that states shall furnish relevant information concerning the objects they have launched into outer space. But how should space objects be registered?

The Registration Convention establishes a double system of registration, either in a national registry,[30] or in the international register held by the U. N. Secretary General.[31] A key difference among the two is that, while the former is only accessible upon permission of the state of registry, the latter is publicly available to all states. This fact makes the international registry held by the U. N. Secretary General a crucial source of information about manmade objects orbiting Earth. According to Article IV of the Convention launching states are requested to furnish information and data to the U. N. Secretary General for insertion in this registry. However, the functionality of this system is affected by two factors: (1) the nature of information provided; (2) the timeframe to provide that information.

As to the first problem, often states are reluctant to disclose the real purpose of the space object, particular if it is of a military nature. As to the second point, the Convention requires states to furnish information to the U. N. Secretary-General "as soon as possible." The terminology "as soon as possible" is vague and leaves states a discretionary choice as to when to provide such information, being before or, most likely, after the launch has taken place.

In recent years the Registration Convention, although being increasingly ratified, has been experiencing some difficulties, mainly connected with the current practices of on-orbit transfer of ownership of space objects and the growing involvement of international organizations in space activities. These difficulties led to discussions on the issues of the 'launching state' and 'registration of space objects' in the Legal Sub-Committee of UNCOPUOS and resulted in the adoption by the U. N. General Assembly of resolution 59/115 on December 2004 and resolution 62/101 on December 17, 2007.

Finally it should be noted that, in addition to the U. N. registration process, there is a parallel process that requires the registration of radio frequency and orbital locations with the International Telecommunication Union (ITU). This is to avoid physical and radio frequency interference and optimize the effective use of particular orbits such as the geosynchronous orbit. Only national administrations of the ITU undertake the registration process, and, thus, international entities such as Intelsat, Inmarsat, and Eutelsat provide their information to a national administration to complete this process (i.e., the United States registers Intelsat information, the United Kingdom registers Inmarsat information and France registers

Eutelsat information). The ITU has a coordination process if there are indications of interference between satellite networks.

The 1979 Moon Agreement

The fifth and final international space treaty is the 1979 Moon Agreement.[32]

Unlike the other space treaties, the Moon Agreement has received a limited number of ratifications and none of the space powers is Party to it. Two main reasons can explain this phenomenon: (1) the controversial nature of some provisions of the Agreement, particularly that introducing the concept of the "common heritage of mankind"[33]; (2) the limited interest of states in accepting binding obligations governing the exploration and use of the Moon and its resources.

Although the first point will be addressed later, as to the second one it should be pointed out that, after the successful manned landing on the Moon in 1969, states lost interest in the Moon and decided to invest their resources on other more practical and useful space applications, such as telecommunications and meteorological satellites. Thus, by the time the final text of the Moon Agreement was ready and open for signature in 1979, the Moon was no longer at the center of the space agenda of states. This situation remained unchanged for over 30 years. However, in recent times a renewed interest in the Moon and its resources has emerged, and several unmanned lunar missions have been launched. This has also led to discussions within UNCOPUOS concerning the status and future of the Moon Agreement with a view to understanding the reasons for its lack of appeal and to increasing the number of parties.

The Moon Agreement establishes rules governing the exploration, use and exploitation of the Moon[34] and its natural resources. The Agreement makes a clear distinction between activities of scientific and non-scientific, i.e., commercial, nature. In the course of scientific activities states may: (a) land their objects and personnel on the Moon (Article 9); (b) establish manned and unmanned stations on the lunar surface (Article 10); and (c) collect and remove samples of lunar minerals, which shall remain at their disposal for the purpose of scientific research (Article 6, para 2). These samples can be made available to other countries upon request. Significantly, the agreement makes clear that the placement of personnel, facilities and stations on the surface of the Moon does not create any right of ownership over the area of the Moon involved in the operation (Article 11, para 2).

Non-scientific activities, in particular the commercial use or exploitation of lunar resources, are regulated under Article 11 of the agreement. According to Article 11, para 1, the Moon and its natural resources are defined as the "common heritage of mankind."

Generally speaking, the concept of the "common heritage of mankind" refers to the common management by nations of international areas containing valuable resources. This common management shall result in the establishment of an international regime, and, theoretically, in the setting up of an international authority, aimed at regulating the removal and commercial use of the area's resources.

Among the most controversial aspects of the 'common heritage of mankind' the following can be listed: (a) the idea that the benefits generated from the area's resources shall be shared equitably among nations, regardless of their level of participation in the exploitation activities; (b) the structure of the decision-making mechanism of the area's international authority; (c) the request of technological assistance from developed to developing countries. Several developed states as well as private entrepreneurs claim that the common heritage of mankind introduces anti-competitive measures and is detrimental from an economic perspective.

Article 11 of the Moon Agreement is, however, far from establishing a clear and comprehensive regulation of the exploitation of lunar resources under the 'common heritage of mankind heading.' The agreement does not establish an international regime to govern such exploitation. Any decision concerning this is postponed until the moment in which the exploitation of lunar resources becomes feasible. The agreement only provides for the general purposes of such a regime including the orderly and safe development of the natural resources of the Moon and the equitable sharing by all states in the benefits from those resources, whereby the interests of all countries as well as those of the countries that have contributed to the exploitation activities shall be taken into account.

Some states and legal analysts have pointed out that the provisions of Article 11, and, in particular, the clause about the sharing of benefits, contribute to create a climate of uncertainty concerning the exploitation of lunar resources and the profits to be generated from it.[35] Thus, these experts have suggested that the "sharing" provisions have a detrimental rather than positive effect on future lunar initiatives. These criticisms appear to be partially unjustified. As described above, the Moon Agreement leaves it to the states to decide how to structure the rules to govern the exploitation of lunar resources; moreover, while requiring the equitable sharing of benefits, it points out that such a sharing shall be done taking into account not only the interest of developing countries but also that of developed ones directly or indirectly involved in the exploitative activities.

Non-binding Instruments

During the 1980s and 1990s, it became evident that it was no longer possible to conclude internationally binding documents regulating space matters; thus, the United Nations realized that the best way to continue developing space law was through declaring legal principles by resolutions of its General Assembly.[36] Consequently, four set of such principles were elaborated in UNCOPUOS and adopted by the General Assembly, namely: the 1982 Direct Television Broadcast Principles[37]; the 1986 Remote Sensing Principles[38]; the 1992 Nuclear Power Sources Principles[39]; and the 1996 Space Benefits[40] Declaration.[41]

Broadly speaking, the U. N. General Assembly Space Resolutions are elaborations of ideas previously laid down in the 1967 Outer Space Treaty or the application of their generalities in particular instances.

Unlike the space treaties, the resolutions adopted by the General Assembly are not legally binding. Nevertheless, the principles that they include are relevant for three reasons: (1) They express a legal opinion of the international community; (2) They may contribute to establish customary rules of international law; and (3) They may constitute the first step towards concluding an international treaty at a later stage.[42] Significantly, evidence shows that states tend, to a relevant extent, to comply with the resolutions' provisions. Legally speaking, the U. N. principles constitute 'soft law' instruments.

Resolution 37/92 on Direct Broadcasting by Satellite

This resolution introduces general regulatory norms of conduct for the use of a special category of space technology, namely the use of satellites for direct broadcasting.

The Direct Broadcasting by Satellite (DBS) principles have been an object of controversy as to their scope, purpose and implementation. These factors have seriously undermined the success of Resolution 37/92, which has remained largely unapplied.

Controversy has centered on the way direct broadcasting services should be provided. Pursuant to the DBS principles any such service shall require the prior consent of the states interested in the transmission of the broadcasted signal. This approach has been criticized for going in the direction of effective control over satellite broadcasting and leaving little room for the concept of free flow of information. In today's era of worldwide communications, it is questionable whether the philosophy chosen by the DBS principle has still any significance; indeed, the idea of free flow of information has gained wide support.

Resolution 41/65 on Remote Sensing of the Earth from Outer Space

Remote sensing means gathering information without physical contact. The remote sensing principles address the sensing of Earth from space for the purpose of improving natural resources management, land use and the protection of the environment. There are several benefits derived from remote sensing activities, in particular that of viewing large sections of Earth at regular intervals.

The remote sensing principles lay down basic rules according to which remote sensing from space should be carried out.[43] The most significant aspect that the principles deal with is the relation between the sensing state (i.e., the state operating a remote sensing satellite) and the sensed state (i.e., the state whose territory is observed from space).[44] Firstly, the sensing state has the right to observe another state's territory from space without the need to request an authorization to do so; in short, there is no rule of prior consent to being sensed. Secondly, the sensed state shall have access to the primary and the processed data of its territory on a 'non-discriminatory basis'; analyzed information of its territory shall also be made available. This does not mean that access to data and information by the sensed state has to be free or automatic, just non-discriminatory and at 'reasonable' cost. Thirdly, even if data are shared, the sensed state needs to have mechanisms in place (expertise, people, hardware) to make use of them.

Significantly, the sharing and selling of remote sensing data and information is regulated by 'data policies.'[45] It should be kept in mind that remote sensing data are the result of investments; thus, nations or companies want to obtain a return from them. This is why data policies, especially for commercial undertakings, establish strict rules on copyrights, marketability and use of remote sensing products. In other cases, the element of 'public benefit' prevails and data policies are more oriented towards open data access and distribution.

The Remote Sensing Principles have found large application in practice; for these reason they might be considered as constituting customary law.

Resolution 47/68 on Principles Relevant to the Use of Nuclear Power Sources (NPS) in Space

As a starting point it shall be pointed out that for certain types of space exploration, such as for solar exploration beyond the orbit of Mars, there is no alternative to nuclear power. Nuclear power sources can be used for two reasons: (1) to provide power (i.e., to run the system of a spacecraft); or (2) to provide propulsion (i.e., to move the spacecraft). The most common use is the provision of power.

These principles regulate the use of nuclear power sources in outer space.[46] Taking into account the risks that the use of nuclear power encompass the principles set forth a series of precautionary measures that space operators need to implement. These measures are intended to protect individuals (on Earth or in space), the biosphere, and the outer space environment.[47]

Resolution 51/112 on Space Benefits[48]

The Space Benefits Declaration addresses one of the key principles of space law, namely the exploration and use of outer space to be carried out for the benefit and in the interests of all countries, with particular regard to the interests of developing countries. The idea for having a Declaration on "space benefits" originated from the complaints of the developing countries that argued that this principle, set forth in Article I, para 1, of the Outer Space Treaty, had not been fully applied by the developed states. The Declaration was intended to provide means and directions on how to implement this principle. However, the Space Benefits Declaration falls short from achieving this result; in fact, it only contains provisions of a general nature that add little to the already existing rules regulating space activities.

The only significant aspect of the Declaration is the fact that it makes clear that international cooperation in space activities shall be based on the free decision of states, i.e., no form of forced cooperation (Para 2), and that such cooperation shall be conducted in the forms and methods that are considered most effective and appropriate by the countries concerned (Para 4).

UNCOPUOS Developments in the Early Twentieth Century

In the past decade, the Legal Sub-Committee of UNCOPUOS undertook a review of the existing body of U. N. space law, aimed at discerning its shortcomings and suggesting possible ways forward.[49] Two issues were deemed worthy of special attention: (a) the concept of 'launching state'; (b) the registration of space objects. Discussions on these two topics resulted in the adoption of two new resolutions by the U. N. General Assembly. These resolutions were not meant to provide an authoritative interpretation of, or proposed amendments to, the Liability and Registration Conventions. Instead, they merely suggested certain practices to ensure a coherent application of these conventions.

The topic of the 'launching state' was selected because it has become increasingly complicated to identify from the perspective of liability. This has been the consequence of developments in launch technology and the growth of commercial opportunities in space, i.e., the possibility of cooperation in launching services. These factors are challenging the traditional notion of 'launching state', according to which a 'launching state' is the state from whose territory or facility the launches takes place, or the state that procures the launch. The identification of the "launching state" is made even more complicated by the fact that launch can take place not only from within territories, but also from the high seas or an aircraft; furthermore, it may occur that the state having control over a satellite under Article VI of the OST and the launching state do not coincide, particularly where a transfer of property in orbit has occurred.

For these reasons, UNCOPUOS wanted to clarify the notion of the launching State. U. N. General Assembly Resolution 59/115 relating to "the application of the concept of launching state" was adopted on December 10, 2004. The Resolution encourages states to undertake various steps in connection with their space activities: firstly, to adopt national space legislation to enable continued supervision of non-governmental entities under their jurisdiction; secondly, to conclude agreements with other launching states in relation to joint launches and other forms of cooperation; and thirdly, to foster consistency of national space legislation with the provisions of international law.

As far as the registration of space objects is concerned, issues have emerged in relation to: (a) the timing and the information provided by states pursuant to the Registration Convention; (b) registration in case of joint launches; and (c) transfer of ownership of satellites in orbit. A working group to address these problems was established within the Legal Sub-Committee of UNCOPUOS. As a result of deliberations within the working group, UNGA Resolution 62/101 on "Recommendations on enhancing the practice of states and international inter-governmental organizations in registering space objects" was endorsed on December 17, 2007.[50]

The resolution makes concrete proposals both to achieve uniformity in the type of information provided to the U. N. Secretary General and in the registration of a space object in case of a joint launch. As to the issue of transfer of ownership in orbit, the resolution does not exclude it. However, it clarifies that such a transfer does not imply a transfer of jurisdiction and control to a non-launching state, which remains within the original launching state. Indeed, the responsible purchasing nations may acquire responsibility and liability through a bilateral agreement with the state of registry without a change in jurisdiction and control under international law. This information may also be registered as a supplementary report in the U. N. Register. However, this does not happen in the majority of the cases. Therefore, in order to enhance transparency of space activities, Resolution 62/101 recommends the state of registry to furnish to the U. N. Secretary-General some additional information, such as the date of change in supervision and the name of the new owner or operator.

Legal Developments Outside of the UNCOPUOS Framework

Preliminary Considerations

In the past decades the space law-making activity within the United Nations and, in particular, within UNCOPUOS, has progressively slowed down. The limited mandate of committee, the impossibility to agree on the insertion of new and timely items in its agenda and political disagreements have been at the root of these problems.

This, however, does not mean that new rules to govern space activities have not been formulated. Instead, norms have been developed through a process involving a variety of new participants and taking place outside of the traditional UNCOPUOS framework, such as in international inter-governmental and non-governmental organizations, as well as on a bilateral and multilateral basis. Often, these norms are inserted in non-binding documents, such as codes of conduct, thus leaving their application to the free will of nations.

Although this approach may appear limited or weak if compared to the traditional one within UNCOPUOS, currently it constitutes the most workable solution for the evolution of space law at the international level. This process is known as the development of 'soft law' as opposed to the traditional 'hard law' as contained in treaties and conventions.

The need for new space law rules is driven by four main factors. Firstly, there have been developments in the fields of science and technology that have contributed to the expansion of the uses and applications of space technologies. Secondly, there are an increasing number of countries capable of launching satellites into orbit. Most of these states are enhancing both their military and civilian space capabilities. Thirdly, we've seen the rise of new commercial space capabilities and activities. This increase in private enterprises involved in space activities is changing the traditional role of governmental entities dominating all aspects of space activities and the closely related areas of space policy and regulation. Fourthly, there has been an emergence of new legal and technical issues that were not foreseen or considered relevant at the time of the drafting of the U. N. space treaties, including the problem of space debris.

The following paragraphs will first describe the development of rules on a bilateral level or within the context of inter-governmental organizations. Then, the analysis will focus on specific issues, such as space debris, financing of space assets, space traffic management, and on the initiatives undertaken to develop legal frameworks to address them.

Bilateral and Multilateral Arrangements

Despite the fact that bilateral and multilateral norms have been developed outside of the UNCOPUOS framework some new legal obligations have emerged. These types of agreements often regulate cooperation between states and governmental space agencies. The United States alone had already concluded, following a count made in the late 1990s, more than 1,000 technical and scientific agreements with some 100 countries and international organizations. This number today must be twice that amount and continues to increase.

Also international inter-governmental organizations, such as the ITU, ESA, etc., play an important regulatory, administrative and legal role at the international level. Indeed, they create not only legal obligations among respective members but they can also address the regulation of specific aspects of space uses and space cooperation.

Of particular significance was the creation of ESA, whose Convention entered into force in 1980,[51] to guide and coordinate the activities of European states in space matters. ESA has gradually become an influential space actor not only in the implementation but also in the progressive development of new norms governing the cooperation of states as well as non-state actors in space matters.

Having been invested by its convention with legal personality,[52] the Agency was also empowered to "cooperate with other international organizations and institutions and with governments, organizations and institutions of non-member nations, and conclude agreements with them to this effect."[53] Accordingly, ESA has concluded several space cooperation agreements with its member countries but also with many non-European governments and governmental space agencies. On one side, such agreements create a stable legal framework for many types of space operations; on the other side, they contribute to the development of space law.

Regulation of the Issues of Space Debris, Financing of Space Assets and Space Traffic Management

The analysis now focuses on three key issues of the current space law debate, namely: (1) the regulation of orbital space debris; (2) the financing of space assets; and (3) space traffic management. Several attempts to formulate rules regulating each one of these issues have been undertaken. These rules have taken different shapes, including norms, standards or other statements of expected behavior in the form of recommendations, charters, guidelines, and codes of conduct.

Space Debris[54]

Attention towards the problem of space debris has significantly grown in recent years. Although there is still no agreed binding definition as to what is a 'space debris,' this term has been increasingly used in deliberation within UNCOPUOS. The term has been applied with reference to "all man-made objects, including fragments and elements thereof, in Earth orbit or re-entering the atmosphere, that are non-functional."[55] Space debris represent a problem because, due to the high velocity (7,500 m/s or much higher) with which these objects move around Earth, they constitute a threat to functional space objects, and if they are large enough or contain hazardous materials can even impose risks when they de-orbit.

The U. N. space treaties do not specifically address space debris, mostly because space debris was not an issue at the time the treaties were negotiated. International discussions on the regulation of space debris started in the early 1980s, but it was only in 1993, upon the initiative of the world's major space agencies, that the Inter-Agency Space Debris Coordination Committee (IADC) was established.[56]

After several years of discussions, the IADC developed space debris mitigation guidelines in 2002. These guidelines, which became the basis for the space debris mitigation guidelines developed and adopted by UNCOPUOS in 2009,[57] establish a series of measures and good practices aimed at reducing the risk of the creation of debris. These guidelines are voluntary in nature and not legally binding under international law. Consequently, no binding international norms regulating space debris exists today. Nevertheless, space agencies have been implementing the guidelines for over a decade. Furthermore, several states have included in their national space legislation provisions on space debris mitigation and prevention. These provisions are obligatory for the actors, both of governmental and non-governmental nature, which have been authorized by those states to carry out activities in outer space. The insertion of these types of provisions in national space laws can be seen as an instrument to transform non-obligatory international norms into rules that are enforceable at least on a national scale (Fig. 1.2).

The United States has the most advanced set of regulations dealing with the environmental aspects of space activities. Already in 1995 NASA had issued a comprehensive set of procedures for limiting orbital debris—the NASA Safety Standard 1740.14. In 1997 the Debris Mitigation Standard Practices, based on the NASA Safety Standard, were developed by NASA and the Department of Defense (DOD). The Debris Mitigation Standard Practices are applicable to space systems (including satellites and launch vehicles) that are "government-operated or produced." These standard practices have four objectives: (1) control of debris release

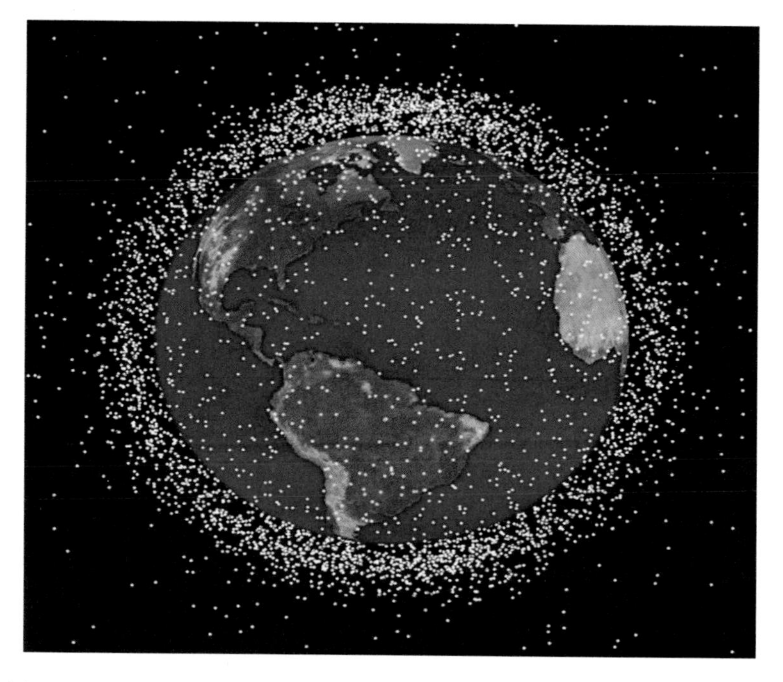

Fig. 1.2 A computer-generated graphic of the population of debris in low earth orbit, a region of space within 2,000 km of Earth's surface. (Courtesy of NASA at http://orbitaldebris.jsc.nasa.gov/photogallery/beehives.html)

during normal operations; (2) minimization of debris generated by accidental explosions; (3) choice of safe flight profile and operational configuration; (4) post-mission disposal of space objects, either by re- or de-orbiting.

In addition, many U. S. governmental agencies have developed guidelines and regulations relating to orbital debris generation and mitigation in their respective areas.[58] For example, the Federal Aviation Administration (FAA), which is the U. S. agency empowered to issue space launch and re-entry licenses both for public and private missions, requires a licensee to receive a safety approval. In order to get such approval, the applicant must demonstrate its ability to comply with specific space debris mitigation and prevention requirements.[59] A significant advancement was achieved in 2004, when the U. S. Federal Communications Commission (FCC), the agency in charge of issuing licenses for radio transmissions by private entities in the United States and, thus, also regulating communications satellites, released a new comprehensive set of rules concerning mitigation of space debris. These rules, which substantially incorporate the IADC guidelines concerning spacecraft disposal, apply to the licensing of commercial U. S. satellites as well as to the use of non-U. S. satellites.

The FCC regulations provide for operational and disclosure requirements in three categories: avoidance of collision with large objects during normal operations, post-missions disposal, and assessments and analyses to prevent a spacecraft from becoming a source of debris. They require U. S.-licensed geostationary orbit (GEO) satellites to be placed at the conclusion of their lifetime into a disposal orbit above the GEO. The commitment to re-orbit satellites pursuant to this rule is a pre-condition for receiving a license. Operators also have the obligation to submit an orbital debris mitigation disclosure, in which they provide information about their plans and practical compliance with the debris mitigation and prevention rules.

Other countries have progressively developed their own debris mitigation guidelines and made them applicable to 'nationals' involved in space activities. For example, the Russian Federation, Japan, France, Italy and the United Kingdom have their own space debris guidelines.[60] Due to the largely congruent interests of space operators in terms of debris mitigation, these guidelines tend to be similar and broadly consistent. Of course, they present some differences, but the fundamental principles are based on the IADC guidelines. In this regard, these national regulations normally address debris mitigation during normal operations, call for depletion or neutralization of a spacecraft at the end of its lifetime, and demand an assessment of collision risks to avoid breakups.

Finally, the European Union (EU) Draft Code of Conduct for Outer Space Activities also addresses the issue of space debris.[61] The Draft Code, an instrument meant to be applicable to the space activities of the subscribing states and non-governmental entities under their jurisdiction, contains several provisions aimed at preventing, reducing and mitigating the creation of space debris.[62]

Yet another aspect of controlling space debris has been the formation of the Space Data Association. This international non-governmental entity, founded by Intelsat, Inmarsat, and SES Global, has now expanded to several dozen spacecraft operators. This body maintains an active database that allows all members to know of possible conjunctions (i.e., collisions) that might occur.[63]

The Financing of Space Assets

Growth in the commercial uses of outer space has also directly influenced the drafting of new rules to govern space activities. In the last decade, thanks to relaxation of government control and an increase in the civilian applications of space technologies, several non-governmental, i.e., private, entities have entered the outer space realm, attracted by the potential economic revenues resulting from its commercial uses.

The presence of private operators creates new legal challenges in the field of space law, especially the need to set forth norms directly applicable to them. Private entities operating on an international scale still face several barriers to their full participation to space ventures. In particular, they still struggle to obtain such financing.

In order to address the latter issue the International Institute for the Unification of Private Law (Unidroit)[64] adopted, after nearly a decade of discussions, the Protocol to the Convention on International Interests in Mobile Equipment on Matters Specific to Space Assets on March 9, 2012. [65]Unidroit is an intergovernmental organization, based in Rome, that is particularly active in the development of modern legal solutions for commercial development—involving all the key parties, both of governmental and non-governmental nature.

The Space Protocol is designed to facilitate the asset-based financing of high-value mobile projects, whether satellites or, in future, space vehicles, especially for those parties most in need of such financing, notably the smaller operators and start-up companies. It is expected that, as with the Protocol to the same convention on matters specific to aircraft equipment (which, although only adopted a decade ago, is already in force in 44 states and already provides the basic legal framework for aviation secured financing), the enhanced legal certainty provided by the new regulations will significantly assist those parties struggling to raise the enormous sums required to acquire, launch and operate satellites and other space assets. Specifically, the Protocol should increase competition in the commercial space market and thus create new opportunities to finance private space business. However, it remains to be seen what will be the reaction of states as well as private operators to its adoption.

Space Traffic Management

The idea of space traffic management is linked with that of space sustainability, a concept that refers to the unimpeded use of outer space for all participants today, as well as in the future, and to an adequate level of safety for space activities. Several space experts insist that the existing space legal framework is unable to meet these needs. Consequently, they suggest establishing mechanisms to regulate 'space traffic.'

The most elaborate proposal is included in a study published in 2006 by the International Academic of Astronautics (IAA).[66] The IAA study points out that, despite the limited dimension of space traffic compared to traffic on Earth, there is a high risk of collision among space objects; this is the consequence of their limited maneuvering capability and their high speed. The study suggests a set of technical and regulatory provisions for promoting safe access, operations, and return from space.

Although the study has not been formally adopted by any state, it has contributed to make them aware of the risks of an unregulated and unresponsive use of outer space. Following the IAA study, other initiatives on the issue of space traffic management have been undertaken. In this respect, the European Code of Conduct for Outer Space Activities stands out. The code aims at setting forth what can be seen as a core for a future comprehensive space traffic management program, as it includes provisions focusing on the safe and responsible access and use of outer space. A workshop on this topic hosted by the International Civil Aviation Organization (ICAO) and the McGill Air and Space Law Institute was held in May 2013 in Montreal, Canada, with co-sponsorship from the Secure World Foundation, the International Association for the Advancement of Space Safety (IAASS) and the International Space Safety Foundation (ISSF). This workshop was intended to explore the practicalities of a space traffic management system and the legal system under which it might operate.

Another relatively new page in the development of space law is national space legislation. This topic will be extensively addressed in the following chapter of this book.

Chapter 2
National Space Legislation

Overview

Space activities have traditionally been limited to governmental entities. In the early decades of the Space Age only states had the technical and financial capabilities to carry out the exploration and use of outer space. Private actors were long excluded from it. On one side, due to strategic and political reasons, governments were hesitant to allow anyone other than their own military or governmental space agencies to be involved in space-related affairs. On the other side, the enormous financial commitments required to undertake any kind of operation in space deterred potential private investors from entering the space sector. Furthermore, the legal framework to regulate private activities in space was largely inadequate.

However, in the 1980s and 1990s private operators started entering the space market. This was mainly due to the need for states to find additional sources to finance space activities. Indeed, already in the late 1970s, governments started reducing their space budgets, and other, preferably private, financial means had to be secured to support space projects.[67] Additionally, technological advancements and the reduction of costs contributed to attract private investors towards outer space.

Initially, only public–private partnerships, for example long-term cooperation between the public and private sectors to execute projects typically in the hands of the public sector, were organized. Later, entirely private undertakings appeared, although up to now they remain a limited number.

Importantly, the privatization of space activities has gone hand in hand with its commercialization. Broadly speaking, the term "commercialization" refers to making money and profits. In the context of space, it can be intended as the use of equipment sent into or through outer space to provide goods or services of commercial value.

The United States was the first country to embrace the commercialization of outer space, mostly after the collapse of communism in the early 1990s. Nowadays, thanks to the increasing civilian application of space technologies and the

F. Tronchetti, *Fundamentals of Space Law and Policy*,
SpringerBriefs in Space Development, DOI: 10.1007/978-1-4614-7870-6_2,

globalization of the economy, the privatization and commercialization of space are becoming key components of modern space activities.

The growing privatization and commercialization of space activities created new legal challenges to space law. In particular, that of adapting a legal framework developed having in mind states as the main space actors. It soon emerged that the formulation of national space legislation was the most suitable instrument to achieve this goal.[68] This choice was largely due to the consideration that international space law requires states to ensure that national space activities are consistent with the fundamental space law principles. Thus, it seemed more logical for each state to individually regulate the involvement of their nationals in space undertakings rather than developing new internationally agreed rules.

Several states have so far enacted national space laws and regulations.[69] These tend to differ in their scope and content. The differences are the consequence of the characteristics, extent and degree of space activities effectively being carried out under the supervision of a particular government.

The following sections are organized as follows. First, the analysis of why national space legislation is enacted and the issues that such legislation normally deal with will be provided. Second, a few examples of national space legislation will be given. And finally, specific areas related to space activities from a national perspective, namely export control, will be addressed.

Enacting National Space Legislation: The Issues

Several reasons can be enumerated to explain why a growing number of states have been enacting national space legislations. First, activities in outer space are inherently hazardous. In order to prevent harm caused by space operations carried out by private entities, a mechanism to supervise and control them is needed. Second, states need to make sure that private subjects, while operating in space, do not violate a nation's international obligations or undermine its national security and foreign policy interests. Third, the international legal regulations for space activities and, in particular, the Outer Space Treaty, the Registration Convention, and the Liability Convention, impose numerous obligations on governments that cannot somehow be transferred to private entities. These obligations make it necessary for countries to adopt national space legislation.

Authorization and Supervision of National Space Activities

Article VI of the Outer Space Treaty is the most relevant provision of the U. N. space treaties concerning the participation of private operators in space activities.[70] It creates a special link between private entities and their states, by making the latter internationally responsible for the space activities of the former.

Although not directly requiring the adoption of national space legislation, its practical implementation has led several states to do so.

The first sentence of Article VI of the Outer Space Treaty reads as follows:

> States Parties to the Treaty shall bear international responsibility for national activities in outer space, including the Moon and other celestial bodies, whether such activities are carried on by governmental agencies or by non-governmental entities, and for assuring that national activities are carried out in conformity with the provisions set forth in the present Treaty.

This sentence makes clear that nations bear international responsibility for national space activities, both carried out by governmental agencies and non-governmental entities. In the context of Article VI, the concept of "responsibility" has a different meaning than the one normally given to it under general international law. Normally, a state bears 'direct responsibility' only for acts somewhat directly attributable to it.[71] Instead, under Article VI, a state is responsible for all space activities carried out by private entities falling under its jurisdiction. Furthermore, states must ensure that private space activities are conducted in compliance with the obligations laid down in the OST.

The second sentence of Article VI explains how states can actually perform this task, as it provides that: "The activities of non-governmental entities in outer space, including the Moon and other celestial bodies, shall require authorization and continuing supervision by the appropriate state Party to the Treaty." Accordingly, states must authorize and continuously supervise the space activities of non-governmental entities. Thus, it is evident that, while private subjects are entitled to carry out activities in outer space, they may only do so if they receive authorization from their state.

To conclude, in order to fulfill their international obligations, states must set forth a mechanism enabling authorization and supervision of non-governmental space activities.[72] Article VI does not directly require the enactment of national space legislation, but ultimately this has emerged as the optimal solution to govern the authorization and supervision of private activities in outer space. Finally, it should be stressed that the scope of national space legislation is not necessarily limited to the implementation of Article VI of the Outer Space Treaty, but it may serve other purposes, such as making sure that private activities comply with safety standards and rules on debris mitigation and prevention as well as ensuring that they do not interfere with security and foreign policy interests of a state.[73]

The Registration Convention

International space law requires countries to register objects launched into outer space at the international and national level. This requirement, first laid down in U. N. General Assembly Resolution 1721 of 1961 and then substantially repeated in Article VIII of the Outer Space Treaty, has been regulated in detail by the 1975 Registration Convention.

The Registration Convention indicates the type of information that shall be provided by a state for its inclusion in the U. N. registry. In order to be able to send the required information, a state needs to put in place a system establishing how and when a space operator will furnish these data. It is only upon receipt of such data that the state is in the position to transmit them to the U. N. Secretary. Normally, such a system is set up in specific provisions of national space law. In addition, a national registry must be established, the contents of which remain at the discretion of the nation. The establishment and functioning of the national registry are also provided for in national space legislation.

The Liability Convention

International space law establishes that a state is internationally liable for damage caused by the objects it has launched in outer space. This principle is laid down in Article VII of the Outer Space Treaty and further elaborated in the Liability Convention.

National space legislation cannot modify in any way how a state's international liability is regulated in the space treaties. It is, however, interesting to see how it organizes the relation between a state and the non-governmental entities in the context of liability.[74] The state that has authorized the space activities of a non-governmental entity remains primarily liable for the damage caused by a space object operated by such an entity. Nevertheless, usually a government establishes a right of recourse against that entity if the former has paid compensation for damage caused by the latter. National space legislation normally incorporates this principle and lays down the prerequisites and conditions according to which the right of recourse is activated and exercised. Furthermore, in order to provide guarantees that the amount paid by the state will eventually be recoverable, national space legislation might force non-governmental entities to obtain insurance covering the launch or the operation of a space object.

In the United States this system has been organized as follows.[75] The issuance and transfer of launch and re-entry licenses for commercial launches is the responsibility of the Office of Commercial Space Transportation of the Federal Aviation Administration (FAA). License applicants, apart from obtaining safety approvals from the FAA, are obliged to take out insurance or otherwise prove their ability to compensate for liability claims brought by third parties or the U. S. government for damage to government property resulting from the licensed activities.

The amount to be covered by the insurance is the "maximum probable loss," which is determined by the FAA. In relation to third party claims, this amount should not exceed $500 million or "the maximum liability insurance available on the world market at a reasonable cost,"[76] while for governmental claims the cap is $100 million or the maximum insurance available at a reasonable cost.[77] In case a third party is successful in bringing a claim that tops the licensee's insurance, the U. S. government will cover the additional amount up to a total claim of $150 million.[78]

This system, based on the limited liability of the licensee, insurance, warranty of the state and "maximum probable loss," has influenced the drafting of several national space legislations adopted in recent years. For example, the French national space act of 2008[79] provides that any operator shall have and maintain, for the period of its space operations, insurance or another form of financial guarantee.[80] The operator will be absolutely liable for damage caused on the surface of Earth or in airspace by a space object connected to an authorized activity.[81] When the French government has paid compensation pursuant to its international liability, it shall have the right to make a claim for indemnification by the operator.[82] However, the latter may benefit from a government guarantee that amounts to € 60 million.[83]

Selected National Space Legislative Actions

The following section gives an overview of some of the most significant national space legislative actions. A comprehensive analysis of all the national space laws adopted up to this moment goes beyond the purpose of this book.

The United States

The United States has the most elaborate national space legislation in the world.

The development of U. S. national space law has been closely linked with the actual conduct of space activities and with technological advancements. Every time technology enabled a certain space activity or the geo-political and economic situation moved in a new direction, a new piece of space legislation was adopted in the United States. This has been the case for the adoption of legislation on private space launches, remote sensing and commercial spaceflights.

In 1958, NASA was established. It was put in charge of running the U. S. civilian space program. This decision was highly significant, as it took the U. S. civilian space program out of the hands of the military. The act that established NASA has been amended several times in order to incorporate elements of contract, tort and insurance law, indemnification and intellectual property.

In 1984, the Commercial Space Launch Act laid down a licensing regimen for private space launches. Its goal was to favor, facilitate and support commercial space launches by the private sector. The act deals with the issuance and transfer of launch and re-entry licenses issued by the FAA. This system, including the related provisions on insurance requirements and compensation for third-party liability, has been previously described, so it is not addressed in detail here.

The 1992 Land Remote Sensing Policy Act regulates the licensing procedures for private remote sensing activities. The act recognizes the fact that in this sector, government funding is still needed and that full commercialization is not reachable in the near future. The act is based on the idea that remote sensing data are important tools for several subjects and purposes and should, therefore, be made available easily and at low cost. On the other hand, it recognizes the commercial value of remote sensing data, by means of a data protection plan, and imposes restrictions based on national security considerations.

The Commercial Space Launch Amendments Act of 2004 introduced a number of provisions aimed at regulating the recent phenomenon of space tourism. The act requires commercial suborbital flight operators to make several written informational disclosures in order to obtain the informed consent of customers, the so-called 'spaceflight participant.' This is a necessary pre-condition to establish that the spaceflight participant accepts the risk inherent in the spaceflight and that he or she is not entitled to the benefit of liability insurance coverage. In 2012 a further amendment was enacted into U. S. law to extend the provisions for experimental licensing arrangements for commercial spaceplane launches.

Apart from these acts specifically addressing space activities, several existing laws from other areas are applicable to space activities. For example, the 1934 Communication Act as amended was declared to be relevant to private operators of space communications activities. Furthermore, there are many regulations that specify more concretely the applicable law to some space activities, such as commercial human spaceflight, remote sensing and U. S. participation in the International Space Station.

As it might appear from the above short overview, U. S. national space law is in fact a rather complex set of laws and regulations applicable to various space activities. In order to make U. S. space law more coherent and easily accessible in 2010 the House of Representatives passed a bill that introduced a new Title 51—"National and Commercial Space Program"—that collects together all existing space laws without modifying them.

The Russian Federation

During the Soviet Union's tenure no dedicated national laws regulating space activities existed. Instead, special resolutions and decisions of government and political bodies constituted the legal basis for operations in outer space. Only after the collapse of the USSR did the process of building up a properly structured national space legislation start in Russia.

The Federal Law on Space Activity of August 20, 1993, is the fundamental document of Russian space law. It gives the right to regulate space activities to the president of the Russian Federation, the Space Agency of the Russian Federation (Roscosmos), and the Ministry of Defense. The Law on Space Activities establishes the procedure governing authorization for all space activities in the Russian

Federation both for scientific and socio-economic purposes. An implementing regulation contains more detailed provisions on the kind, issuing, validity, suspension, and withholding of space licenses. Roscosmos and the Ministry of Defense are empowered to ensure that authorized activities comply with safety standards and do not constitute a threat to the environment.

With regard to liability, the Law on Space Activities sets forth a system of compulsory and voluntary insurance. Space operators are under an obligation to obtain insurance to cover damage to the health and life of cosmonauts, space infrastructure personnel and liability for damage caused to the life, health, or property of third parties. Voluntary insurance may be obtained to cover damage to space equipment and the risk of loss or damage to it.

With respect to registration, any Russian organization that is exploiting a space object or is conducting or procuring its launch must provide Roscosmos with related information one month prior to the actual launch. Seven days following the launch, information about the launched space objects must be transmitted to Roscosmos. After enlisting in the national registry, Roscosmos passes the information to the Ministry of Foreign Affairs, which communicates it to the U. N. Secretary General for inclusion in the U. N. Register.

France

Despite being one of the major space-faring countries France did not have national space legislation for a long time.[84] Until June 2008, French space activities were regulated by general civil, administrative and criminal law, and by specific laws applicable to certain activities, such as broadcasting and telecommunication.[85] In June 2008 France passed the French Law on Space Operations, which sets out a regimen regulating authorization and control of space operations pursuant to France's international commitments.[86] In particular, the act: (a) establishes under which conditions a private operator may obtain an authorization to carry out space activities; (b) separates liability between the state and non-governmental entities; and (c) foresees sanctions to be imposed in case of non-compliance with authorization requirements.

A private entity is granted authorization to perform space activities only after the relevant state authority has reviewed its moral, financial, and professional qualifications and verified its compliance with safety, health and environmental standards. Authorization will not be given if the planned activity is likely to compromise national security and France's international obligations. Furthermore, authorized operators are obliged to comply with debris mitigation and prevention guidelines. The provisions concerning liability and insurance requirements have been previously described, so they are not addressed here.[87]

The national registry of space objects is kept by the Centre National d'Ėtudes Spatialies (CNES). The procedures governing the functioning of this registry are laid down in an implementing decree.

The French Space Act was certainly inspired by the U. S. Commercial Space Launch Act, but it integrates it with some European peculiarities. For example, its scope and application are broader than the U. S. model and covers not only launches but also other kinds of space activities as well as registration. In this sense, the French Space Act may become a model for other European countries willing to develop a national space act.

Belgium and The Netherlands

Belgium and the Netherlands are two countries that, despite not being space-faring ones, have recently adopted national space laws. This choice has been determined, on one hand, by the increased number of space activities by their nationals as well as from their territories, on the other hand by the need to properly regulate and supervise these activities and ensure compliance with international obligations.

The Belgian Law on Activities of Launching, Flight Operations or Guidance of Space Objects was enacted on June 28, 2005. The Belgian Space Act establishes a legal regimen to authorize and supervise space activities performed under Belgian jurisdiction,[88] the creation of a national registry for space objects,[89] and the avoidance of liability that may arise pursuant to Article VI of the Outer Space Treaty.[90] The act has been supplemented by a Royal Decree implementing certain provisions of the law, which was adopted March 19, 2008.[91]

The Dutch Law Incorporating Rules Concerning Space Activities and the Establishment of a Registry for Space Objects was adopted by the Dutch Parliament on January 25, 2007. The Dutch Space Law regulates registration, authorization, and supervision, and gives the possibility of redress in cases of government liability for damage caused by space activities. It imposes compulsory insurance as a necessary condition for obtaining a license.[92] The amount of the insurance is to be determined by the Minister of Economic Affairs, who, in making this decision, will take into account the maximum possible cover for liability arising from space activities and the sum likely to be covered by insurance.[93] If the state is obliged to pay compensation under Article VII of the Outer Space Treaty or the Liability Convention, it is entitled to recover the sum, in full or in part, from the party whose space activity has caused the damage.[94] The licensed operator shall cover the damage caused by its space object only up to the sum insured.[95]

The Belgian and Dutch examples show that the importance of national space legislation also for smaller countries with a limited space industry. As a consequence of the privatization and commercialization of space activities, these states are progressively entering the space arena and need a solid legal framework to accommodate such a development.

Export Control

All countries protect their investments and their intellectual properties with the goal of maximizing their return on these investments. However, this does not mean that technology is not transferred from one entity to another. Such a transfer can also involve sensitive exports and goods. This type of transfer raises serious political and strategic issues, particularly when the technology is used by the military. In this cases, governments exercise two kinds of control: (1) they protect specific technology or knowledge; and (2) they prohibit the export of certain knowledge and/or technologies to specific state.

Technology can be described as industrial know-how, resulting from the combination of knowledge, human resources and technical management, necessary to develop new equipment or expertise for certain purposes. Inventions are protected through a series of governmental measures, such as patents, licenses, and trademarks, that are usually referred to as intellectual property rights.

The U. S. Approach

In the United States,[96] sensitive international technology transfer is controlled through a strict licensing process. International Traffic in Arms Regulations (ITAR) is the name for a set of U. S. government regulations that control the export and import of defense-related articles and services that are included in the U. S. Munitions List (USML).[97] ITARs are interpreted and enforced by the Department of State Directorate of Defense Trade Controls (DDTC).

All U. S. manufacturers, exporters, and brokers of defense articles, defense services, or related technical data, as defined on the USML, must register with the U. S. Department of State. Through this registration the U. S. government obtains information on who is involved in manufacturing and exporting activities. Registration does not confer any export rights or privileges but is a precondition for the issuance of any license or other approval for export.

Pursuant to ITAR regulations information and material concerning defense and military-related technologies (for items listed in the U. S. Munitions List) must only be shared with and sold to Americans, unless authorization from the Department of State is granted or a special exemption is used. Under ITAR, a "U. S. person" who wants to export USML items to a "foreign person" must obtain authorization from the U. S. Department of State before the export can take place.

ITAR also includes clauses on retransfer of USML items by a foreign persons. Accordingly, the "Retransfer" (also called "Re-export") of items on the USML by foreign persons is prohibited unless specific authorization to do so is given under the relevant export authorization.

U. S. persons (including organizations) can face heavy fines if they have, without authorization or the use of an exemption, provided foreign (non-U. S.)

persons with access to ITAR-protected defense articles, services or technical data. Significantly, since 1999 the U. S. government has substantially increased action against organizations and individuals responsible for breaches of ITAR. The most notable enforcement action was the $100 million penalty applied to ITT Corporation as a result of the unauthorized Retransfer of night vision technology to the People's Republic of China in 2007.

ITAR has a huge impact on U. S. space activities and, specifically, on the international trade of space technologies and commodities; indeed, since 1999[98] all satellites, launch vehicles and related items have been transferred to the U. S. Munitions List.[99] Consequently, all commercial communication satellites (comsats) are now subject to ITAR.

One might wonder what are the economic implications of such a situation. It has to be kept in mind that commercial communications satellites are an important source of profit for the U. S. space industry. This type of satellite requires a large capital investment up front that is shared between manufacturers, launchers, and insurers. Commercial satellite operators choose their launch service providers on the basis of technical capability, cost, reliability, insurance rates, and scheduling. All things being equal, a commercial satellite operator not subject to contracting restrictions will launch on the least expensive vehicle.

The U. S. satellite industry argues that ITARs negatively affect its interests; in particular, it slows down and complicate a U. S. satellite company's plans to deliver parts or a complete spacecraft to a foreign customer. Furthermore, ITAR makes it very difficult, expensive and time-consuming for foreign entities to buy U. S.-manufactured satellites and components.

Thus ITAR is accused of sharply cutting into the U. S.'s share of the satellite manufacturing business worldwide. A study by the FAA demonstrates that in the period from 1999 to 2006 a substantial decline in the manufacturing of launch vehicles, satellites and ground equipment, has occurred. For several years, the satellite industry has been working to convince federal lawmakers that the executive branch, not Congress, should have the discretion to decide how to administer satellite exports and that, eventually, ITAR restrictions on commercial satellites should be removed. Interestingly, in December 2011 the House of Representatives passed a bill, H.R. 3288, with a clause to remove ITAR restrictions on commercial satellites. The specific wording for the purpose of the bill: "To authorize the President to remove commercial satellites and related components from the United States Munitions List subject to certain restrictions, and for other purposes." It remains to be said what will be the follow up to the adoption of this bill.

The European Export Control Regulation

European policy on export control[100] is based on the distinction between export items in either conventional armaments (e.g., munitions) or dual-use goods. Conventional armaments lay outside the purview of the European Union, and member states may exempt the manufacturing and trade of arms from the rules of the common market.[101] In short, "each Member State sets up its own policy and procedures for the export of conventional arms." On the contrary, dual-use goods fall within EU purview; consequently, the EU is empowered to oblige all member states to request licenses "to export the items on the list and to have appropriate penalties for violations as well as effective systems for enforcing the relevant legislation."[102]

Such empowerment is given to the EU by Council Regulation No.428/2009.[103] This regulation provides for a common community export licensing system, a control list, and a general export authorization procedure.

Dual-use items are cataloged in Annex I and Annex IV of Council Regulation No. 428/2009. All dual-use items listed in Annex I require export authorization for export from the EU. Annex IV lists items that are deemed to be so sensitive that they need authorization even before they are transferred from one EU state to another.

Comsats and their associated components are categorized as dual-use items under Annex 1. Thus, the export of comsats needs authorizations from the EU.[104] Such authorization is given by the appropriate authorities of the member state where the exporter is located.[105] Exporters are requested to provide the authorities with all relevant information needed for their applications for individual and global export authorization. In particular, information is required on the end user, the country of destination and the end-use of the item exported. Clearly, when deciding about granting an export authorization, member states take into account several factors including the obligations arising from relevant international non-proliferation regulations and export control arrangements, and considerations of national foreign and security policy.

Chapter 3
Global Administration of Outer Space

Overview

Space law has progressively become an area of great interest not only from an academic point of view but also from a diplomatic and technical perspective. This interest is revealed by the increasing number of organizations, bodies, agencies, institutions and associations dealing with space matters.

The above heterogeneous entities can be divided into two groups: (a) those operating within the U. N. framework; and (b) those acting outside of the institutionalized context of the United Nations.

There are obvious differences between these two groups. The first one mostly consists of organizations and agencies of a governmental nature. Their activities are carried out at state/diplomatic level and their decisions/acts might have, depending on the circumstances, a binding or non-binding character. The second group includes non-governmental organizations, the activities of which generally involve technical and scientific experts. The work undertaken in the context of these non-governmental organizations normally results in documents of an optional nature. Despite their distinctions all these entities contribute to the advancement of space law and to the diffusion of knowledge about the legal and technical issues relating to space activities.

The Role of the United Nations and Its Specialized Agencies

The U. N. Committee on the Peaceful Uses of Outer Space

At the dawn of the Space Age states, and in particular the United States and the Soviet Union, were aware of the fact that space activities could offer numerous opportunities for development. At the same time, it was also clear that the use of outer space could alter the strategic international balance and undermine national security. It is thus not surprising that, in the late 1950s, the United Nations was

F. Tronchetti, *Fundamentals of Space Law and Policy*,
SpringerBriefs in Space Development, DOI: 10.1007/978-1-4614-7870-6_3,

chosen by its member states as the most appropriate theatre to discuss: (a) questions relating to the exploration and use of outer space; and (b) the development of a legal framework regulating space activities. At that time several issues remained uncertain: How could military confrontation between the West and the East be avoided in space? Which laws would be applicable to space activities? How could countries lacking the technology to carry out activities in space also benefit from them?

It was under these premises that the General Assembly of the United Nations established the Committee on the Peaceful Uses of Outer Space (UNCOPUOS), first as an ad hoc committee in 1958 and then as a permanent committee of the General Assembly in 1959. UNCOPUOS was given the task to "review, as appropriate, the area of international co-operation, and to study practical and feasible means for giving effect to programs in the peaceful uses of outer space which could appropriately be undertaken under United Nations auspices"[11] and to "study the nature of legal problems which may arise from the exploration of outer space."[12]

Subsequently, the committee established two sub-committees, a Legal Subcommittee (LSC) and a Technical Subcommittee (TSC).[108] The committee and its cubcommittees meet annually in Vienna, Austria, and are assisted by the U. N. Office for Outer Space Affairs (UNOOSA). UNCOPUOS began its activities with eighteen members. Its membership has increased to sixty-nine, a number that supposedly ensures equitable representation of interests and geographical coverage.

UNCOPUOS played and still plays a fundamental role in the development of international space law. The five U. N. space treaties and well as the UNGA resolutions on space have all been negotiated and drafted within this committee. Significantly, UNCOPUOS works through the practice of consensus. Consensus is a form of agreement reached without a vote, which does not necessarily means that there is unanimity on the issue at stake. Abstention from the discussion of a point is not taken to imply dissent, and an individualistic interpretation of particular language may therefore be passed over, unnoticed by, or even concealed from, other parties. Within UNCOPUS a proposed text, whether in the form of a draft treaty, resolution or other statement, is negotiated and revised until all members are willing to accept it and support it.

Consensus presents advantages and disadvantages. As to the former, consensus facilitates compromise and, in the case of draft treaties, the parties may be more keen in due course to ratify provisions in whose drafting they have participated. As to the latter, consensus may result in ambiguous or vague language, leading to diverging interpretations and applications of a certain provision.

Outer space law-making within the United Nations works as follows. After agreeing on the insertion of a certain topic in the committee's agenda, discussions within the Legal Subcommittee of UNCOPUOS begin. After a text has been adopted by consensus by the LSC, it is sent to the main committee of UNCOPUOS. If the committee agrees upon this text by consensus, the U. N. General Assembly adopts a resolution containing this text. If it is a treaty, states can then decide

Fig. 3.1 The annual sessions of UNCOPUOS are held at the U. N. Office in Vienna, Austria, which also hosts the U. N. Office for Outer Space Affairs (UNOOSA). Courtesy of UNOOSA at www.oosa.univienna.org

whether to sign, ratify, or accede to it. If the agreed text is a resolution of the General Assembly, states may decide to apply it, without the need to formally ratify it (Fig. 3.1).

Agencies Within the United Nations

Within the United Nations system a number of agencies deal with space-related matters. These agencies do not exclusively address space issues, but are involved, to a certain extent, in space activities as part of their responsibilities.

The International Telecommunication Union

The International Telecommunication Union (ITU) specializes in telecommunication and aims at: (a) spreading the benefits of information and communication technologies around the globe; (b) enabling the growth and sustained development of telecommunications and information networks; and (c) facilitating universal access to information and communication technology. The ITU works as a center to mobilize technical and financial resources to make it possible to achieve these goals.

The ITU is divided in several sectors, each dealing with a specific issue. The ITU Standardization Sector (ITU-S) develops standards necessary to create infrastructures to deliver telecommunication services on a worldwide basis. The ITU Development Sector (ITU-D) helps countries trying to pursue telecommunication development strategies. The ITU Radio Communication Sector (ITU-R) plays a vital role from a space law perspective, because it provides for equitable management of the radio-frequency spectrum and satellite orbits. These are two limited natural resources that are increasingly in demand for a large number of services (e.g., fixed, mobile, broadcasting, space research, meteorology, global positioning, etc.).[109]

ITU-R's main goal is to ensure equitable and efficient use of radio communications systems through the implementation of Radio Regulations and Regional Arrangements. Such instruments are regularly updated in World and Regional Radio Communication conferences. ITU-R manages the coordination and recording procedures for space systems and Earth stations and examines frequency assignment notices submitted by administrations for inclusion in the formal coordination procedures or recording in the Master International Frequency Register.

Basically, every satellite operator must apply to the ITU-R for an orbital slot in which to place its satellite(s) and for a frequency in which to operate its signal. After reviewing the application, ITU-R may decide to grant the operator the requested slot and frequency and to confer upon the applicant the right to exclusively use the slot for a specific time (normally the operational time of the satellite) and to operate its frequency free from interference by other operators. Recently, the ITU system for the assignment of orbital slot has been facing a problem, notably that of "paper satellites." In short, this problem consists of a growing number of slot applications submitted to the ITU to accommodate the needs of systems that will never leave Earth or are merely speculative. The ITU counteracts such practices by obliging the applicants to provide detailed plans of their intended space activities and to set out a deadline for the deployment of the satellite in the assigned orbital position.

Apart from the ITU, other U. N. agencies involved in outer space matters are worth being mentioned. For example, there is the U. N. Educational, Scientific and Cultural Organization (UNESCO), which deals with science education and the use of space for cultural applications, such as for the monitoring of historical sites; there is also the World Meteorological Organization (WMO), which coordinates worldwide activities related to space-based meteorology. Interestingly, the ITU, UNESCO and the WMO have been participating in the UNCOPUOS sessions since its establishment in 1958.

With the deployment of commercial telecommunications satellites and the launch of civilian Earth observation satellites, some additional U. N. entities started using space technologies to support their operations. For example, the U. N. Environment Program (UNEP) and the Food and Agriculture Organization (FAO) utilize Earth observation data to evaluate the status of Earth's environment and to check food security. Basically, space-based information systems are crucial to the work of the United Nations. In order to ensure coherence and coordination in the uses of space-related products, representatives of the U. N. entities meet annually during the sessions of the Inter-Agency Meeting on Outer Space Activities.

It is also worth mentioning the contribution given to space law by another entity within the United Nations, namely the U. N. Institute for Disarmament Research (UNIDIR).[110] UNIDIR is a voluntarily funded autonomous institute that offers an independent forum for the international community to discuss and promote initiatives relating to the issues of disarmament and security. UNIDIR has been particularly active in the field of space security and prevention of weaponization of outer space. Increasingly, it has emerged as the preferential forum to discuss space security-related matters at the international level, due to the reluctance of several delegations to address similar issues in UNCOPUOS.[111] For example, UNIDIR was chosen by China and Russia as the forum to present their draft treaty on the Prevention of an Arms Space Race in 2008.

International Space-Related Organizations Outside of U. N. Systems

Outside of the United Nations framework various space-related international organizations operate. These organizations are open only to sovereign states; hence, they are commonly referred to as inter-governmental organizations. The number of these organizations has steadily increased, and some of them have permanent status in UNCOPUOS.

The International Telecommunication Satellite Consortium (INTELSAT) was established on the basis of agreements signed by governments and operating entities in 1962, pursuant to General Assembly Resolution 1721 (XVI) that requested "communication by means of satellite should be available to the states of the world as soon as practicable on a global and non-discriminatory basis." Since the launch of the first commercial geo-synchronous communications satellite, called Early Bird (Intelsat I), in June 1965, INTELSAT has been furnishing global satellite telecommunications services. In 2001, following a trend among space operators, INTELSAT was transformed into a private company, Intelsat, Ltd. The inter-governmental responsibilities of INTELSAT were moved to the International Telecommunication Satellite Organization (ITSO), whose goal is to make sure that Intelsat, Ltd., provides public telecommunications services on a global and non-discriminatory basis. Currently, 150 states are active users of the Intelsat network, but it no longer functions as an international inter-governmental organization.

The International Maritime Satellite Organization (INMARSAT), which was founded in 1979 to provide the maritime community with a satellite communication network, has undergone similar changes. It was privatized in 1999 and divided into two entities—the commercial company, Inmarsat plc, and the inter-governmental organization International Mobile Satellite Organization (IMSO), which operates as a supervisory body to guarantee the provision of services through the Inmarsat satellites.

The Intersputnik International Organization of Space Communication (Intersputnik) was established in 1971 in Moscow by the Soviet Union together with a group of socialist states as a means to respond to the establishment of INTELSAT. Nowadays, it operates as an inter-governmental organization with 25 members and as a commercially aligned Intersputnik Holding Ltd. The legal basis of its operation is to be found in the Operating Agreement that entered into force in February 2003.

The final organization of this kind is the European Telecommunication Satellite Organization (EUTELSAT). EUTELSAT was created in 1977 as an inter-governmental organization to provide satellite services for the European market and was turned into a private company called EUTELSAT S.A. in 2001. As in the case of INTELSAT and INMARSAT, EUTELSAT also maintains an inter-governmental body acting as a supervisory authority. Nowadays, EUTELSAT operates on a global basis and ranks among the top three largest worldwide satellite operators.

Apart from these organizations operating on a global basis over the years several space-related organizations and forums acting on a more limited scale have been established. Although their scope is limited, these organizations and mechanisms contribute to international and inter-regional cooperation in the field of space. Of the original communications satellite organizations that were established as international inter-governmental organizations only the Arabsat continues to operate on this basis.

The European Space Agency (ESA) was created in 1975, replacing the European Launcher Development Organization (ELDO) and the European Space Research Organization (ESRO), both of which were established a decade before. According to the ESA's Convention, the purpose of the organization is "to provide for, and to promote, for exclusive peaceful purposes, cooperation among European states in space research and technology and their space application, with a view to their being used for scientific purposes and for operational space application system."[112] As of 2012, 19 states are members of ESA. Canada is associated under a Cooperation Agreement and some East European countries are participating in the Plan for European Cooperating States (PECS), which is effectively an intermediary stage before full membership.

ESA's activities, which are financed by a budget of nearly 4 billion Euros, are divided into two categories: (1) mandatory programs, dealing with basic activities, such as general studies, education, shared technical investments, and (2) optional programs, such as human spaceflight and exploration, Earth observation, satellite telecommunication programs and launcher developers. Member states contribute to the mandatory programs based on their Gross National Product (GNP), while the contribution to the optional programs are left to the discretion of each member. Among ESA's successes are the development of five types of launchers, the design and operation of more than 60 satellites and more than 200 launches.

With the entry into force of the Treaty of Lisbon in 2009 the European Union is given a limited shared competence in space and thus it has been added as a new player, alongside the European governments, ESA and EUMETSAT, to the European space sector. The planned European Global Navigation Satellite System (Galileo), and the Global Monitoring for Environment and Security (GMES) initiative are joint undertakings by ESA and the European Union. This is complicated by the fact that the memberships of ESA and the European Union, although similar, includes different member states.

The interest in space-related matters is also growing in the Asian-Pacific regions. China, India, and Japan represent key examples of states that are becoming prominent actors in on the space scene in Asia.

Two regional Asian organizations have been established to provide a forum for regional space cooperation. These organizations cannot be compared to ESA in terms of scope, activities undertaken and budget available. However, they play an important role in enhancing mutual understanding between the participating countries, spreading knowledge and education of space law, and enabling the realization of common projects, for example in the use of satellite data for disaster management activities. The two organizations are: the Asia Pacific Space Cooperation Organization (APSCO) and the Asia–Pacific Regional Space Agency Forum (APRSAF). APSCO,[113] which held its first meeting in 2008, currently consists of nine nations that have signed the APSCO Convention, namely Bangladesh, China, Indonesia, Iran, Mongolia, Pakistan, Peru, Thailand and Turkey. Among its main technical initiatives are atmospheric research, spatial data sharing, the development of the Asia–Pacific Ground Based Optical Space Objects Observation System (APOSOS) and the APSCO Applied High Resolution Satellite System, an optional program.

APRSAF was established in 1993 in response to a 1992 declaration adopted by the Asia–Pacific International Space Year Conference (APIC) to improve space activities in the Asia–Pacific region.[114] Participating organizations, including space agencies, governmental bodies, international organizations, companies, universities and research institutes, meet annually under the framework of APRSAF in a hosting Asian-Pacific country. APRSAF is the largest space-related conference in the Asia–Pacific region. The two most visible achievements of APRSAF are the Sentinel Asia initiative for disaster management[115] and the Satellite Technology for the Asia–Pacific Region (STAR) program for capacity building through the development of small satellites.

The Role of Non-Governmental Organizations

Several organizations and working groups dealing with space-related issues have been established in the past decades. These range from academic institutions and formal conferences to international non-governmental organizations. The purpose of these organizations may change over time. However, they usually perform one of the following functions: (a) to try to enhance international cooperation in the field of space activities; (b) to offer a specialized environment for discussing and spreading knowledge about issues related to space law and activities; or (c) to contribute to the development of space law through the drafting of legal documents.

International Association of the Advancement of Space Safety

Safety is the condition of being safe and free from danger.[116] The idea of "space safety" refers to the absence or mitigation of risks associated with civilian uses of outer space.[117] Due to the increasing number of active objects in orbit and the presence of myriads pieces of space debris, the possibility of collision and accidents in space has significantly increased. Ensuring safety in outer space has thus become a priory concern. States as well as technical and legal experts have undertaken efforts to develop a set of standards and rules of behavior regulating civil uses of space.

The International Association for the Advancement of Space Safety (IAASS),[118] legally established on April 16, 2004, in the Netherlands, is a non-profit organization dedicated to furthering international cooperation and scientific advancement in the field of space systems safety. The IAASS membership is open to anyone having a professional interest in space safety.

The goals of the organization are:

1. Advance the science and application of space safety.
2. Improve the communication, dissemination of knowledge and cooperation between interested groups and individuals in this field and related fields.
3. Improve understanding and awareness of the space safety discipline.
4. Promote and improve the development of space safety professionals and standards.
5. Advocate the establishment of safety laws, rules, and regulatory bodies at national and international levels for the civilian use of space.

The IAASS dedicates most of its activities to doing research and providing publications in the field of space safety. The IAASS research initiatives include issuing general studies in support of global safety risk management and standardization, as well as of detailed studies on specific topics. Among the most recent topics addressed under the umbrella of the IAASS, two can be highlighted: (1) the development of space safety regulations and standards; and (2) the need for an integrated regulatory regimen for air and space in the light of the emerging private commercial spaceflight business. [119] An important characteristic of the IAASS's research activity is that it places emphasis on the involvement of academia and representative of the industry. In this way, the IAASS hopes to get the broader possible support to its initiatives.

International Space Safety Foundation

The sister organization of the IAASS is the International Space Safety Foundation (ISSF) that was established in the United States in 2008. This foundation, together with the IAASS, sponsors the publishing of the *Space Safety* magazine (www.spacesafetymagazine.com). This website and magazine not only provide updates on technology and operational information related to space safety but also provides information with regard to space safety regulations, standards, and legal issues that arise with regard to this field.[120]

Space Law Institutions

The International Astronautical Federation (IAF) is a worldwide federation of organizations involved in space.[121] The IAF, which was established in 1950, serves a major hub for discussions relating to the exploration and use of outer space. It annually organizes a global space conference, the contents of which are published in the *Acta Astronautica* and the *Proceedings of the International Institute of Space Law*. The IAF remains to this day the only international federation for the space community that addresses all aspects of space—developments, activities, knowledge, experts and the future. The IAF is now the world's leading space advocacy organization, with 226 members in 59 countries including all leading space agencies, space companies, societies, associations and institutes (Fig. 3.2).

In the 1960s the IAF established the International Academy of Astronautics (IAA), membership of which is open to individuals active in all forms of space activities. Its function is to gather together individuals to exchange ideas and studies and, thus, to contribute to the advancement of space and astronautics. In 1958 the IAF decided to establish a Permanent Legal Committee, open to all jurists affiliated with the IAF, in order to study the legal problems relating to the exploration and use of outer space. The name of this committee was changed in 1959 into the International Institute of Space Law (IISL). The IISL, which is now one of the leading non-governmental bodies dealing with space law, holds annual colloquia during the congress of the International Astronautical Federation. The proceedings of these colloquia constitute an important contribution to the corpus of space law.

The other major non-governmental organization active in the field of space law is the International Law Association (ILA).[122] This association, which was founded in 1873, is open to any lawyers active in space law. Through the work of

Fig. 3.2 Picture taken at the 2012 International Astronautical Congress held in Naples, Italy. (Used by permission of the International Astronautical Federation)

its branches and international committees, the ILA studies and helps clarify international law. Its Space Law Committee has formulated a number of reports on space law covering issues such as the revision of the U. N. space law treaties and questions concerning the commercial use of space.

Chapter 4
Dispute Settlement in Outer Space

Overview

International relations are inextricably linked to the emerging and settling of disputes. Although international law is created to balance diverging interests and to prevent international disagreements, it is nearly inevitable that disputes connected to the interpretation and application of international rules arise. In order to maintain international peace and stability and to ensure the correct application of the law, international law has long established methods and practices to settle international disputes, the so-called dispute settlement mechanisms.

The importance of having reliable and efficient international dispute settlement mechanisms has significantly grown. This has been the direct consequence of the augmented level of international cooperation in various sectors of the economy, the increased number of subjects involved, and the need for creating a stable and predictable environment to carry out business activities. Consequently, various mechanisms to settle specific kinds of disputes have been established.[123]

Despite having an international nature and being characterized by international cooperation, until very recently outer space law was characterized by a nearly total absence of specific mechanisms to settle outer space-related disputes. Such an absence was the consequence of the fact that space activities were the exclusive realm of a limited number of states, and it was believed that recourse to traditional means to settle disputes available under international law, in particular bilateral discussions, was sufficient. In particular, no specialized settlement mechanisms for outer space were thought to be needed.

The situation is different today. First of all, the number of space-related activities has augmented, primarily due to an increase in the commercial uses of outer space in sectors such as satellite communications, launching services, and remote sensing. Secondly, the number and types of actors involved in space activities, now including states, inter-governmental organizations and private entities, has arisen. It is thus reasonable to assume that, due to these factors, outer space-related disputes might occur on an increasing basis.[124]

F. Tronchetti, *Fundamentals of Space Law and Policy*,
SpringerBriefs in Space Development, DOI: 10.1007/978-1-4614-7870-6_4,

As far as dispute settlement is concerned, the existing legal framework regulating space activities presents a major issue—the absence of a compulsory dispute settlement machinery accessible to all space actors. The seriousness of this problem is amplified by the fact that the traditional means for dispute settlements in international law as well as the few available within outer space law, are limited in their personal and material scope and are normally not accessible to private parties.[125] Consequently, private operators are substantially left with no means to settle their international space-related disputes. This situation contributes to a climate of uncertainty potentially discouraging to private investors and companies interested in being involved in space activities.

In the past few years, the international community has increasingly become aware of these problems and has launched initiatives aimed at providing space law with a modern and efficient dispute settlement mechanism. In this respect, the most significant achievement has been the adoption by the Permanent Court of Arbitration (PCA) of a set of Optional Rules for Arbitration of Disputes Relating to Space Activities. Although it is too early to evaluate the impact of these rules, their adoption certainly represents a step in the right direction.

The following sections will be divided in two parts: the first part will give a general overview of the issue of dispute settlement in outer space law. The second part will focus on the need to provide space law with a dedicated dispute settlement mechanism. In this regard, special attention will be dedicated to the most significant initiatives and proposals, such as the 1998 ILA Taipei Draft Convention and the 2011 PCA Arbitration Rules for Outer Space Disputes.

Assessment of Existing Procedures

International space law includes only a few procedures for the settlement of disputes, none of them having a mandatory nature. The 1967 Outer Space Treaty, the 1968 Rescue Agreements, the 1975 Registration Convention and the 1979 Moon Agreement do not contain any specific provision for the solution of conflicts. Only the 1972 Liability Convention lays down a procedure to obtain compensation for damage caused by space objects, which may also include the setting up of a Claims Commission. However, accessibility to these mechanisms is restricted to their members and in relation to cases connected to the purpose of such organizations.

The Outer Space Treaty

The 1967 Outer Space Treaty lacks dedicated provisions on dispute settlement; instead, it includes clauses aimed at minimizing the possibility of conflict among parties. For example, the treaty declares outer space free for "exploration and use by all States"[126] and prohibits national appropriation of outer space.[127] Particularly important for the purpose of conflict avoidance is Article IX of the treaty, according to which if a state has reason to believe that a planned space activity may cause potentially harmful interference with the activities of other states it shall undertake appropriate international consultations before proceedings with such planned activity.

The most significant provision of the Outer Space Treaty concerning the settlement of outer space disputes is included in its Article III, which makes international law and, in particular the U. N. Charter, applicable to space activities.[128] This means that all the dispute settlement means admitted by general international law and the Charter are applicable to activities related to outer space.[129] In this respect, Article 33 of the U. N. Charter contains a non-exhaustive list of such means, including negotiation, enquiry, mediation, conciliation and good offices, arbitration and adjudication.

The means listed in Article 33 can be divided into two groups. The first group, which includes negotiation, enquiry, mediation, conciliation and good offices, comprises non-binding dispute settlement mechanisms. Their characteristic is to result in the adoption of a non-binding act or proposal. The second group consists of binding third-party dispute settlement mechanisms, namely arbitration and adjudication. Recourse to these mechanisms lead to a legally binding solution of an international dispute. Such a binding solution can only be attained when the parties has agreed to confer jurisdiction to settle their case to a third party, either a permanent judicial body or an arbitral tribunal, and to consider as final the decision of this body.

It is important to keep in mind that dispute settlement in international law is consensual by nature. Unless agreed otherwise, no party to an international dispute can be obliged to make use of a specific dispute settlement mechanism. Significantly, states are normally reticent to submit their disputes to third-party dispute settlement machinery. For example, the potential role of the International Court of Justice as a forum for the settlement of space-related disputes is seriously undermined by the fact that none of the space-faring state has accepted the compulsory jurisdiction of the court. Moreover, international law means for dispute settlement are normally available only to recognized subjects of international law, namely states and international inter-governmental organizations. Private entities, lacking the status of subjects of public international law, very rarely are given legal standing before international tribunals.[130] These factors have a negative impact when one considers the possibility of using such means to settle outer space disputes.

The 1972 Liability Convention

The 1972 Liability Convention is the only U. N. space treaty that establishes a specific mechanism for the settlement of disputes arising from the application of its provisions, specifically for disputes concerning compensation for damage caused by a space object in orbit, on the surface of the Earth or to an aircraft in flight. The establishment of such a mechanism was seen by the drafters of the convention as a necessary pre-condition for the fulfillment of its purposes, in particular to ensure the prompt payment of compensation to the damaged parties.

Basically, the convention provides that when parties are unable to settle a dispute through negotiation, they shall set up a Claims Commission. Significantly, the decisions of the Claims Commission will be final and binding only if the parties have so agreed; otherwise, they will merely make recommendations. The Liability Convention contains detailed provisions covering the different phases of the procedure for the settlement of disputes, including the presentation of claims, negotiation, establishment of the Claim Commission, and final decision by the commission.

Liability for damage caused by a space object falls on the launching state. Consequently, claims for compensation for damage shall be presented to the launching state through diplomatic channels.[131] If the claimant state does not maintain diplomatic relations with the launching state it may request another state to present its claim or the U. N. Secretary-General.

Compensation shall be determined in accordance with international law and the principles of justice and equity and shall be sufficient to restore the situation that would have existed if the damage had not occurred.[132] The basis for determining the due amount of compensation has been understood to be the law of the state where the damage occurred.[133]

If no settlement of a claim can be reached through diplomatic negotiations within one year of notification, procedures will continue via the establishment of a Claims Commission under Article XIV. The convention subsequently provides the constitution of the commission in Articles XV–XX. In this respect, the commission shall be composed of three members, one elected by the claimant state, one by the launching state, and the third by the chairman of the commission.[134] Upon establishment, the commission shall determine the merits of the disputes and decide upon the amount of compensation payable.[135]

Highly significant is the provision concerning the binding force of the result of the dispute settlement procedure. According to Article XIX, para 2, the "The decision of the Commission shall be final and binding if the parties have so agreed; otherwise the Commission shall render a final and recommendatory award, which the parties shall consider in good faith. The Commission shall state the reasons for its decision or award." This means that the ultimate effect of the Claims Commission procedure depends on the parties' will. They can agree voluntarily to consider the final decision as legally binding or they can accept it as a recommended opinion. However, in the latter case its effect is again dependent on the free will of the states involved.

Importantly, the dispute settlement mechanism created under the Liability Convention and, in particular, the right of establishing a Claims Commission, has never been used. This does not mean that disputes relating to damage caused by space objects have not occurred; instead, it signifies that state as well as private operators have preferred to settle their disputes directly, i.e., through bilateral discussion. For example, even the most famous case of a satellite crashing on the

surface of Earth and causing damage not to individual persons
environment, namely the Cosmos 954 case, was settled bilaterally, 'ou
by the two states involved in this case, the Soviet Union and Canada, which were both parties to the Liability Convention at that time.[136]

Furthermore, the Liability Convention addresses the standing of international inter-governmental organizations. In case of joint and several liabilities with states, international inter-governmental organizations will bear a preferential responsibility during a period of 6 months.

In conclusion, the Liability Convention is very significant because it lays down a detailed and elaborated mechanism to settle international disputes arising from damage caused by space objects. However, several shortcomings affect its functionality. The major deficiency lies in the fact that decisions of the Claims Commission shall be final and binding only if parties to a dispute have so agreed. In short, the convention does not set forth a compulsory dispute settlement mechanism. Additionally, such a mechanism is limited in its material and personal scope. As to the former, it covers only claims for compensation for damage caused by a space object. Any other claim emerging in relation to space activities falls outside its scope. As to the latter, while this dispute settlement mechanism is directly accessible to states and international inter-governmental organizations, private entities, due to the fact that they cannot be parties to the convention, are not given the same rights and, consequently, are not entitled to directly submit claims to a launching state. A private entity must be able to find a state willing to act in its behalf and to submit its claim. This leaves the initiation of any such action to the discretion of the relevant state with the consequence that, ultimately, private entities may be precluded access. Finally there is the issue noted earlier that the Liability Convention as currently in force is perceived by many as serving as a hindrance to active debris removal.

International Inter-governmental Organizations and Bilateral Agreements

As previously described, the past decades have seen a proliferation of international inter-governmental organizations. Normally, organizations characterized by a broad scope and a wide membership provide themselves with dispute settlement mechanisms to manage disputes arising among their members. Space law does not represent an exception. Legal procedures for the settlement of disputes has been established in various such international and regional organizations as INTELSAT, EUMETSAT, the ITU, ARABSAT, and ESA. The means of dispute settlement provided by the constitutions of these organizations vary from negotiations to compulsory third-party arbitration. The choice of dispute settlement mechanism generally depends on the structure of the organization and the identity of the parties involved in the dispute.

Recent Developments

In the early years of space activities the issue of the establishment of a specialized mechanism to settle outer space-related disputes remained mostly confined to academic circles.[137] States did not consider such an establishment to be a priority and deemed recourse to the traditional public international means for the settlement of international disputes to be sufficient.

Nowadays, the situation is different. As said previously, not only has the number of space actors increased but also their nature has changed. Thanks to the commercialization of space activities, private entities have entered the space realm and have become significant players. Considering the enlarged number and the variety of space operators, along with the growing financial value of space activities, it is reasonable to expect an increase in space-related international disputes.

The previously described limited material and personal scope of the dispute settlement procedures available in space law have been increasingly recognized by states as well as experts as a serious gap in modern space law. This lack of an effective dispute settlement process can actually reduce the participation of subjects interested in carrying out space activities, especially in the case of private operators. Consequently, the need for establishing a specialized mechanism to settle disputes relating to outer space activities has arisen. An improvement in liability processes would also be beneficial.

The choice of the appropriate dispute settlement mechanism for space is not an easy one. Space law was created as a branch of international law, and it is based on public international law principles. However, the growing commercialization of space activities and the resulting involvement of private operators belonging to different jurisdictions, have contaminated space law with elements of private international law. Therefore, a growing tension between public and private international law has become evident. International arbitration and, in particular, international commercial arbitration, has been identified as the preferential method for the settlement of outer space disputes, mostly because it is accessible to all type of entities, both of governmental and non-governmental nature, and because it leaves parties with the choice of applicable law and arbitrators.

The 1998 ILA Draft Convention

The ILA embarked on a study of the issue of dispute settlement in space law as early as 1978. In 1984 the first Draft Convention on the Settlement of Space Law disputes was elaborated. This draft underwent several changes and revisions until 1998, when the "Final Draft of the Revised Convention on the Settlement of Disputes Related to Space Activities" (hereafter the 1998 ILA Draft Convention) was adopted during the meeting held in Taipei.[138]

The basic idea behind the 1998 ILA Draft Convention is to establish an obligation to settle disputes, coupled with the free choice of means and timing for each phase of the procedure to prevent disputes lasting an undefined period of time. Any decision of the dispute settlement body selected under the convention shall be final and binding.

The 1998 convention is applicable to all activities in outer space and activities with effects in outer space.[139] The convention is open to states and intergovernmental organizations. Private entities can make use of the convention through the states that have authorized their space activities.

The convention provides non-binding[140] and binding[141] settlement procedures. The former consist of the obligation to exchange views,[142] and any other peaceful means of dispute settlement.[143] The latter are to be initiated at the request of any party to the dispute when no settlement has been reached following recourse to the non-binding procedures.[144] They offer a choice of means without any hierarchical structure, namely[145]: (1) the International Tribunal for Space Law, if and when such a tribunal has been established; (2) the International Court of Justice; or (3) an arbitral tribunal, constituted in accordance with the provisions of the convention.

The choice of procedure can be made when parties sign, ratify or accede to the Final Draft Convention by means of a declaration.[146] A party, which is party to a dispute not covered by a declaration in force, shall be considered to have chosen arbitration as a method to settle its dispute.[147]

The 1998 ILA Draft Convention has not been signed or ratified by any state. In order to understand this negative response, it should be pointed out that the convention presents some shortcomings. For example, it should have given more weight to issues of accessibility and standing for individuals and small commercial enterprises engaged in space activities. It should have also foreseen some means of universal applicability instead of resorting to the traditional country and intergovernmental organization dichotomy. Furthermore, the approach chosen by the Draft Convention with regard to the possibility of selecting different binding means for dispute settlement appears, nowadays, a bit outdated. Especially, the establishment of an International Tribunal for Space Law and submission of cases to the International Court of Justice do not seem viable options.

The PCA Rules

On December 6, 2011, the Administrative Council of the Permanent Court of Arbitration (PCA) adopted the Optional Rules for Arbitration of Disputes Relating to Outer Space Activities (Outer Space Optional Rules). This adoption represented an attempt to provide space law with a means of voluntary and binding dispute resolution available to all parties engaged in outer space activities and structured to match the particularities of this unique area of economic activity.[148] The Outer Space Optional Rules were the outcome of 2 years of work, started in November 2009 and concluded in December 2011, by a group of experts who operated alongside the International Bureau of the PCA.

Evolution of the PCA Rules

The Permanent Court of Arbitration is an international organization comprised of 115 states.[149] One of its main functions is to facilitate dispute resolution, including arbitration between states, state entities, inter-governmental organizations, and private parties. In this respect, the PCA's secretariat, the International Bureau, offers full registry services and legal and administrative support to arbitral tribunals and commissions.

Significantly, since 1992 the PCA has adopted eight sets of sector-specific rules of procedure for arbitration or conciliation developed by expert groups. In 2009, taking these experiences into account, the Administrative Council of the PCA approved the establishment of an Advisory Group of legal experts ("Advisory Group"), with a mandate to: "...firstly...assess generally the need for a final and binding dispute-settlement mechanism for disputes involving the use of outer space by states, inter-governmental organizations and private entities and, specifically...highlight the benefits of arbitration in this regard. Secondly, the Advisory Group will draw up optional rules to this end for inclusion in the PCA's set of arbitration rules."

Upon establishment, the Advisory Group started addressing the first parts of its mandate, namely to consider the need for a final and binding dispute settlement mechanism for outer space disputes. The Advisory Group pointed out that, in light of the increased number of space activities, particularly as a result of the commercialization of space and the variety of space actors, the potential for disputes had increased. This risk is further augmented by the high level of financial and technical investments involved. The Advisory Group also stressed the significance of the various gaps in the existing dispute settlement mechanisms in international space law. A special problem concerns private entities, as they are precluded access to many of these mechanisms. Private parties may be inclined to resort to international commercial arbitration. At the moment, international space law arbitration agreements between private parties generally provide for arbitration under the U. N. Commission on International Trade Law Arbitration Rules (UNCITRAL Rules)[150] or the procedural rules of private arbitration institutions. These rules, however, praised for being applicable to "the circumstances of various types of disputes and procedures," are not necessarily adapted to space-related disputes.[151] Considering the above, the Advisory Group concluded that an effective dispute settlement mechanism in space law was needed.

Having examined several options, the group considered arbitration as the most suitable dispute settlement mechanism to cope with the current characteristics of space law. The choice was justified on the basis of the following points: (1) Arbitration is a method of dispute resolution open to all parties involved in space activities; (2) Arbitration is a voluntary mechanism, based only on the consent of the parties. This consent can be expressed by inserting an arbitration clause in the legal instrument that defines the parties' relationship. In space law, this instrument can be an inter-state treaty, an agreement between a State and the space industry, a commercial space contract between private enterprises or a private enterprise and a State agency.

The voluntary—or "optional"—nature of arbitration is important in space, as states may be more willing to agree to binding dispute resolution under discrete agreements than to enter into a new significant multilateral treaty, or to establish a specialized court or an additional chamber of the ICJ. Third, arbitration results in final and binding decisions. In arbitration, no appeals are possible and only limited grounds for challenge are available. This can be highly significant because space activities often operate on strict schedules, especially as regards the time windows for landing, atmospheric re-entry, descent and landing, and orbital insertion. Fourth, arbitral awards are internationally recognized and enforceable in all signatory states of the New York Convention, presently 146.[152] Fifth, arbitral procedure is flexible and can be modified by the parties. This is an important factor due to the rapidly evolving character of space activities. Sixth, parties to arbitration choose their own arbitrators. Finally, arbitration can preserve the confidentiality of sensitive information. Confidentiality is crucial in the field of space law due to the economic and security interests involved.

Then, the Advisory Group moved to its second task, the drafting of optional rules for arbitration of disputes relating to outer space. The work of the group resulted in a preliminary Draft of the Optional Rules sent to states in May 2011. Following replies the draft was modified and later adopted by consensus by the PCA Administrative Council.

The Outer Space Optional Rules are largely based on the 2010 UNICITRAL rules and on a set of PCA procedural rules.[153] The UNCITRAL rules are the most widely used set of procedural rules in international commercial arbitration. They were considered an attractive model by the Advisory Group because since the adoption of their first version in 1976, a substantial amount of case law is available. Furthermore, it was deemed that the Outer Space Optional Rules should benefit from the lesson of the first PCA's set of sector specific rules, the PCA Optional Rules for Arbitration of Disputes Relating to Natural Resources and/or the Environment. These rules, which aim at enabling the resolution of disputes concerning the environment and natural resources, shares traits with outer space activities, including a high level of technical complexity and the sensitivity or confidentiality of information pertinent to the arbitral proceedings.

Overview of the PCA Outer Space Optional Rules

Among the most significant elements of the PCA Outer Space Optional Rules the following can be mentioned:

Applicability: The scope of application of the Optional Rules is extremely broad. All those that may be involved in (commercial) space activities, including states, inter-governmental organizations, non-governmental organizations, corporations and private entities, are entitled to use the rules. The applicability of the rules is furthermore extended by the fact that "the characterization of the dispute as relating to outer space is not a necessary pre-condition for the settlement of such dispute under the Rules." When drafting the Optional Rules, the Advisory Group concluded that the geographic, technological or other factual particularities of the dispute should not frustrate the parties' stated intent to proceed to arbitration. The *ratione materiae* jurisdiction of the Outer Space Rules, thus, depends solely on the will of the parties and in no way on any conception of "outer space."

Arbitration Clause and Immunity Issues: The wide applicability of the Outer Space Rules is also reflected in the number of ways in which space disputes can be handled, namely rule, decision, agreement, contract, convention, treaty, or constituent instrument of an organization or agency. These disputes can be settled under the rules upon insertion of an arbitration clause in their text. Significantly, when states agree to insert such a clause, they also agree to waive any right of immunity to jurisdiction.[154] Annexed to the text of the rules there are a model arbitration clauses for contracts and a possible waiver statement.

List of Arbitrators and Experts: Parties to a dispute submitted under the rules are free to select a panel composed by one, three or five persons/arbitrators. These persons can be chosen from a list of knowledgeable persons. This list is provided for by the PCA. [155]Furthermore, the arbitral tribunal can appoint some technical and scientific experts in the field of space law and technology. The PCA provides the list of such experts.[156]

Assistance: The PCA acts as registry and furnishes secretariat services for proceedings under the rules, assisting the parties in choosing arbitrators and performing other legal and administrative functions.

Mechanisms to Avoid Unnecessary Delays: The rules include several provisions aimed at preventing unnecessary delays caused, for example, by obstructive practices of one or more parties[157] or by the failure or incapacity of one of the arbitrators to perform his or her duties. These provisions are intended to guarantee that the arbitration is conducted at a certain speed and concluded within a reasonable period of time.

Confidentiality: Parties are entitled to request the arbitral tribunal to maintain the confidentiality of the final award and the documents circulated during the arbitral proceedings.[158] Confidentiality is extremely important in the field of space law, where disputes involve technical, scientific or commercially sensitive issues. Significantly, Article 17(8) provides for the appointment of a "confidentiality adviser," whose role is to report to the tribunal on an issue containing confidential information, without revealing the confidential content of the document to the tribunal or the other party.

Legal Value of the Award: The award of the arbitral tribunal, made in writing, is final and binding on the parties. Parties are under the obligation to apply the arbitral awards without delays. The final and binding nature of the award is highly relevant in space law because it can contribute to a climate of legal certainty necessary for the pursuit of commercial activities in outer space.

In conclusion, the Optional Rules constitute a significant step forward in the development of space law and a valid basis for the settlement of outer space-related disputes. Clearly, their success depends on the confidence that they will be able to generate in space operators. However, some of their key elements, i.e., confidentiality, broad accessibility, optional nature and binding value of the award, make it uncertain whether the rules will be positively received by outer space operators.

Part II
Space Policy

Chapter 5
Space Policy: The Political and Strategic Impact of Space Activities

Overview

Outer space has progressively become a central component of international relations. Thanks to technological developments, regulatory changes, and a favorable political environment, outer space has acquired a growing significance both from an economic and a strategic point of view.

In terms of its economic impact, in the year 2011 the space sector grossed total commercial revenues of $110.53 billion, including profits deriving from the use of commercial satellite services in the areas of telecommunications, Earth observation and positioning services.[159] Furthermore, the total institutional spending on space in 2011 was approximately $72.77 billion. This figure includes $44.92 billion of civilian expenditures (or 61.7 % of the total) and $27.85 billion of defense expenditures (38.3 %).

As far as the military applications of space are concerned, satellites are largely used to support military operations on the ground; with regard to its civilian uses of outer space, satellites contribute to enhance human activities in several areas, such as resource management, environmental protection, climate change, etc.

Space activities are, thus, a mix of civil, commercial, and military interests. Balancing these diverse elements may be, at times, a quite challenging task. Practice shows that international cooperation constitutes an excellent tool to maximize the benefits deriving from space applications on a global scale.

Nowadays, international cooperation in space-related projects or application is a widespread trend. However, the size and extent of such cooperative efforts is directly affected by the dynamic of international relations. In particular, changes in the international political and economic environment boost or restrict the dimension and nature of space activities. On the other hand, space activities also shape international relations to a certain extent. For example, space assets are indispensable tools used by governments to tackle a number of transnational challenges, such as global warming, the fight against terrorism, the proliferation of weapons of mass destruction (WMD), and economic development issues.

F. Tronchetti, *Fundamentals of Space Law and Policy*,
SpringerBriefs in Space Development, DOI: 10.1007/978-1-4614-7870-6_5,

Considering the strategic significance of access and use of outer space, it is thus not surprising that the technologically advanced nations have begun formulating national space policies. These policies, which may take several forms, including formal national space policies as well as political strategies, indicate the goals of the national space program, the approach towards international cooperation, and the guidelines to be followed by governmental and non-governmental national subjects involved in space activities.

In order to give a short but useful description of the issue of space policy, the following sections will first describe the political and economic dimension of space activities. Then there is a summary of the most significant aspects of the space policies of some space-faring states. This chapter will conclude by addressing two key topics affecting national and international space relations, namely space security and the commercialization of outer space.

The Economic Dimension of Space Activities

The economic significance of space activities has increased in the last decades. Nowadays, the development of new space applications is largely associated with their inherent commercial potential.[160]

During the Cold War, due to national security concerns and the huge risks associated with investments in space, the space sector was government-driven. At that time, space was not seen as an economic sector as such, but rather as a limited scientific and technological domain. Furthermore, states were both the dominant actors and the only clients for space products and services. Because of these factors, serious commercial space endeavors began only in the 1980s. These early commercial activities concentrated on launch services and technological spinoffs. However, the transfer of technology from public to commercial applications still represented a marginal phenomenon, both in terms of size and revenues. Similarly, the launch market did not seem to have strong economic potential, due to a limited market and the influence of governments.

The end of Cold War created a more favorable political and economic environment that enabled a paradigm change in the space economy. Political decision-makers, particularly in Europe, became increasingly aware of the potential of new space applications. The technology-push logic that prevailed in Cold War times was gradually substituted by a new demand-pull approach. Several factors contributed to this development.

First we have seen the emergence of the phenomenon known as globalization. This has brought about waves of liberalization and privatization in the economic markets. The space sector was also affected by this trend, and two acts played a central role:[161] the WTO Agreement on Basic Telecommunications (1997), which led to liberalization of the telecommunications market, and the U. S. Orbit Act of 2000, which promoted competitive global satellite communications.

Secondly, national space budgets have been reduced. This, in a way, forced private firms that fueled the governmental demand for hardware and services to look for new markets. Eventually, this gap was filled through an increase in private sales.

Thirdly, we have seen new levels of cooperation among the United States, Europe, and Russia on several projects. This was the case in government-driven areas, such as those related to space exploration and space science, with the International Space Station as the leading example of cooperation of this kind. Cooperation endeavors took place also in commercial areas, especially in launch service markets. Many joint ventures were created to provide launch services, such as the Russian American International Launch Services (ILS) offering Proton commercial launches. Finally, there was a rise of new applications with high economic impact in the fields of Earth observation, satellite navigation, and satellite telecommunications.

As a result of these factors new actors emerged on the commercial scene and international commercial competition increased drastically both in the upstream and in the downstream sectors. The share of commercial activities in the global space economy has grown steadily since the beginning of the 1990s. Today, commercial space activities represents the biggest share of space economies. Furthermore, new wave of mergers and consolidations in the space markets occurred. While in the 1990s over 20 companies were involved in the design and producing of aerospace systems, today Boeing and Lockheed Martin are the two giants in the upstream markets. A similar trend occurred in Europe where, through mergers and acquisitions, the European Aeronautic Defense and Space Company N.V. (EADS) created the biggest aerospace group worldwide.

Overview of Global Space Economy

After having exposed the main components of space economy, the following paragraphs will give an overview of the main space markets. This description is intended to give an idea of market sizes, players involved, and ongoing trends.

Upstream Sector The main activities forming the upstream sector are the launch market, satellite manufacturing, ground equipment and the insurance market.

According to a report of the FAA, the commercial launch revenues grew steadily between 2004 and 2008, witnessing an increase from roughly $1 billion in 2004 to almost $2 billion in 2008. Also the year 2009 recorded a 26 % increase compared to 2008.[162] However, in 2011 only 18 commercial launches took place.[163] Although U. S. companies did not conduct any commercial launches, Russian companies held the biggest share with a total of 10 launches. Moreover, there were 4 European launches, 2 commercial launches operated by China and 2 additional from the multinational Sea Launch AG. The revenue from the 18 launches amounted to an estimated $1.9 billion, a decrease of 21.7 %, or $ 526 million, from the previous year. European revenue was once again the highest, reaching $880 million, followed by Russian commercial launch revenue at approximately $707 million.

Satellite manufacturing represents the most profitable market in the space sector. Revenues in the years 2009, 2010, and 2011 remained substantially stable, grossing an average of $ 15 billion yearly. The biggest players are U. S. and European companies: Thales Alenia Space and EADS Astrium for Europe; Space Systems/Loral, Lockheed Martin, Boeing, and Orbital Sciences for the United States. New actors are also emerging in this sector, such as India and China, as the Indian Space Agency (ISRO) and China Great Wall Corporation (CGWC) have entered the manufacturing business.

Ground equipment revenues include infrastructure elements, such as mobile terminals, gateways and control stations, and consumer equipment, such as very small aperture terminals (VSAT), ultra small aperture terminals (USAT), DTH broadcast dishes, satellite phones and digital audio radio satellite (DARS) equipments. Portable navigation devices (PND) form one of the sub-segments of end-user electronics idncorporating GPS chip sets. Ground equipment is a market in expansion and in 2010 represented the 31 % of the world space business revenues.

Finally, insurance costs have constantly decreased, as the space industry has demonstrated high reliability and low rates of accidents. The total coverage value of the 175 satellites currently insured is approximately of $17 billion dollars.

Downstream Sector The downstream sector includes satellite services, which is the area generating the highest revenues in the space economy. The three major space applications are satellite communication (DBS, FSS and MSS), remote sensing and navigation.

In 2011 and 2012 the satellite services industry maintained its growth despite operating in adverse economic conditions. Enlarging demand from emerging economies and developing regions allowed an expansion of satellite capacity and revenues. Furthermore, this industry continues to expand, demonstrating an ability to adequately mix investments in innovation technology and new services while reinforcing operating functions.

In 2010 satellite services' revenues exceeded $101.3 billion, due mostly to an increase in direct-to-home (DTH) satellite services.

In conclusion, nowadays one should not underestimate the economic impact of space activities. Indeed, space is not only a dynamic and fast growing economic sector, creating wealth and jobs, but it is a strategic sector associated with very high R&D costs and investments. This unique combination makes outer space a relevant factor influencing economic strategies and trends on a global scale.

The Political Dimension of Space Activities

Due to the wide range of space applications addressing Earthly problems in several areas, it is rather evident that space activities have a political dimension.[164] Ensuring the free access and use of outer space as well as the security of space objects constitute key goals of the most developed countries and influence their choices on a global scale.

The Cold War Times

During the Cold War, space activities were directly connected with politics, both from a national and international perspective. Space programs pursued political objectives and mirrored and implemented political priorities. From 1957 to the end of the 1980s, the space arena was largely dominated by the two superpowers, the United States and the Soviet Union. In this period, access to and use of outer space became a foreign policy tool, and space technologies were mostly developed to fulfill military and security goals. Furthermore, national security concerns directly shaped space programs. In the Soviet Union, the development of ballistic missiles capable of delivering nuclear weapons, which was meant to reduce the gap with the United States, could also be used to pursue the goals of the Soviet space programs. Similarly, in the United States, the launch of Sputnik-1 in 1957 generated a sense of urgency and strategic weakness. Indeed, the rocket used to launch Sputnik-1 could also carry nuclear warheads into U. S. territory. This represented a crucial reason to boost the development of rocketry in the United States.

Post Cold War Era to the Present

The political significance of outer space greatly increased after the end of the Cold War. Firstly, outer space started being recognized as a means to achieve several political objectives, not only in the military field but also in economic, social, and environmental matters.[165] Because space science and technologies could now be used to solve concrete domestic and transnational problems, space acquired a broader political scope. Satellite applications could indeed be employed in various areas such as security, transport, navigation, environmental monitoring, weather forecast, rescue management, etc.

The second factor that influenced the political and economic growth of outer space was the increasing number of states involved in space activities. Geopolitical and technological factors enabled this development. In particular, the end of the division of the world into two blocks created political room for smaller states to build their own space capabilities. In addition, the emergence of new space applications, enabling socio-economic benefits, made space attractive to a number of countries, especially in the developing world. Furthermore, thanks to the wave of liberalization and the easement of economic restrictions that followed the conclusion of the Cold War, private entities began entering the space market attracted by its possible profits.

Summing up, nowadays, the use of outer space and of its application has become central not only to pursue strategic goals but also to improve the social and economic conditions of people all over the world. Consequently, outer space issues influence political decision-making both in the developed and developing world.

Chapter 6
Strategic, Economic and Political Space Policies and Issues

After having discussed the political and economic significance of space activities, this chapter gives some examples of national space policies. Furthermore, it addresses some key issues having special significance on the strategic, economic and political dimension of space activities.

National Space Policies: A Few Examples

The United States

The first space policy document enacted by the United States was the 1958 Space Act that established NASA.[166] NASA was made responsible for the coordination of space activities. Among its goals, NASA had to focus on Earth and space science and on the development of launch vehicles and satellites. A specific political goal was to preserve U. S. leadership as a space-faring state.

Two space policy reviews took place under the Presidents Carter[167] and Reagan,[168] stressing the relationship between the civilian and military aspects and the central role of the space shuttle in U. S. space policy. An important space policy document was the 1996 National Space Policy released under the Clinton Administration.[169] This policy mirrored the changes that occurred after the end of the Cold War and mentioned national security, foreign policy, economic growth, and commercial space activities as main priorities.

The 1996 policy was replaced by the 2006 U. S. Space Policy.[170] The 2006 policy took a stronger approach towards national security by emphasizing that, in order to "preserve its rights, capabilities, and freedom of action in space," the United States would "deny, if necessary, adversaries the use of space capabilities hostile to U. S. national interests." This constituted an important doctrinal shift from 'space control' to 'space dominance.'

The 2006 National Space Policy was ultimately replaced by the 2010 National Space Policy, adopted during the current Obama Administration.[171] The 2010

F. Tronchetti, *Fundamentals of Space Law and Policy*,
SpringerBriefs in Space Development, DOI: 10.1007/978-1-4614-7870-6_6,

policy softened the tone of its predecessor and put greater emphasis on international cooperation and transparency and confidence-building measures to avoid conflicts in space. Significantly, it puts particular emphasis in promoting and supporting a competitive U. S. commercial space sector, which is considered vital to the continued progress in space.

Europe

European efforts to shape a coherent space policy represents a prominent example of creating a political framework for space, and taking into account the full potential of space as a socio-economic and political tool. This has been the result of a long and progressive policy process, which culminated which the adoption of the European Space Policy (ESP) in 2007.

European decision-makers recognized very early the potential contribution of space to European politics.[172] Strong political character was given to the process leading to the ESP, as the EU progressively took the lead in shaping the European Space Policy agenda. ESA, the leading player on the European space science scene, had achieved impressive scientific and technological successes, but, as an inter-governmental institution, it did not have the political authority to develop an ESP by itself.

The policy process leading to the adoption of the ESP started in 1999, when the European ministers asked the European Commission (EC) and ESA to build a European Space Strategy (ESS).[173] The ESS was laid down in an EC communication entitled "Europe and space: turning to a new chapter" in September 2000.[174] Three objectives were identified: strengthening the foundation for space activities (mainly access to space and developing a technological and industrial base); enhancing scientific knowledge; and reaping the benefits for markets and society.

The Green Paper program, a series of EC/ESA joint workshops in 2003, and the publication of the subsequent White Paper, were the next steps. After the creation of specialized institutional bodies to deal with space matters, the Space Council, the High Level Space Policy Group (HSPG) and the ESA/EC Joint Secretariat, the ESP was finally adopted in May 2007.[175]

As a whole, the ESP is a very comprehensive framework. It focuses on the development of applications, on security and defense, the maintenance of strong foundations in space with a special focus on access to space, an emphasis on science and technology and exploration, and support for a competitive industry policy. Also European national space actors have developed their national space programs and national space law. With the entry into force of the 2009 Treaty of Lisbon a "space competence" of the European Union has been added to the already complicated legal European "spacecape."[176] The European Union has not been given exclusive competence over European space matters, but a "parallel competence," which means that individual European Union's member states retain sovereign discretion to draft and implement their own national policies and legislations in this area.

The Russian Federation

After the collapse of the Soviet Union, Russia re-organized its space activities and redefined its space policy. The adoption of the 1993 Law of the Russian Federation About Space Activity was the seminal moment of this process.[177] The law emphasizes that space applications and science aims at solving socio-economic, scientific, technical and defense tasks of the Russian Federation. The law created the Russian Space Agency and established that the Federal Space Program (FSP) should lay down the long-term Russian strategy in space.[178] In this respect, Russian space activities cover virtually all possible areas: development of space applications in the areas of Earth observation, Global Navigation Satellite System (GNSS) and satellite communications, space and Earth science, manned flights, defense, commercial activities on the launch market and participation in international ventures.

Japan

Japan began its involvement in space activities in the early 1950s. Thanks to U. S. support, it developed its own launch vehicles and satellites. However, due to a series of failures affecting its launch vehicles and satellites, Japanese space policy underwent a crisis. This led to a reconstruction of the Japanese space sector, which resulted in the creation of the Japanese Space Agency (JAXA) in 2003. A major step forward was the endorsement of the Basic Space Law in 2008, which opened the way for the Basic Space Plan of June 2009.[179] This document suggested an increase in funding and stressed the need for military space applications, taking into particular account the growing security concerns of Japan in the region. Furthermore, it called for a political shift towards needs-oriented applications, such as climate monitoring.

China

The establishment of the Chinese space program was motivated partly by political considerations and heavily by military ones. The development of strategic missiles capable of carrying nuclear warheads was seen as crucial against the threats posed by the United States and the USSR. In the 1980s the emphasis of the Chinese space program shifted from military considerations to the development of satellite applications and commercial activities, such as launch services.

In the 1990s a new strategy oriented towards international prestige was added. This resulted in ambitious program, including the 2003 first Chinese space manned mission and the Chinese lunar program. Currently, China has a broad space portfolio, which is comprised of a fleet of launch vehicles, launching civilian and military satellites in the areas of Earth observation, GNSS (Beidou system) and satellite communications, and manned and unmanned space exploration programs. China still lacks a comprehensive and national space policy. Political reasons and the complex mechanism regulating the law-making process in China are at the root of this situation. Chinese space activities are regulated through rules laid down in White Papers prepared by the State Council, which is the chief political body in China. The latest version of this document was issued in 2011.[180]

Other National Space Policies

Apart for space-faring states, several countries from the African and Asian regions started being somewhat involved in space activities, attracted by the socio-economic benefits deriving from the uses of space. For example, many African countries have understood the advantages of using space applications to achieve the Millenium Developments Goals, especially for environmental monitoring, resource management, and social development. Consequently, several small space programs, mostly in the field of Earth observation, have been launched. The three most active actors in the African continent are Nigeria, South Africa, and Algeria. In Asia, apart from the extensively developed Indian space program, several other actors are showing increased space ambitions. For example, South Korea has set forth the goal of developing an indigenous launch vehicle by 2017.

A final expression of the political relevance of space is the growing number of cooperative endeavors in space, especially at a bilateral level.[181] From a bilateral perspective, several cooperation agreements cover the areas of space science, exploration, and, most recently, space applications. These agreements involved the main space agencies, i.e., NASA, ESA, JAXA, but also countries from emerging space markets in Africa, Latin America and Asia.

Space Security

Space security is generally understood as being related to the absence of unjustifiable man-made or natural threats to space assets.

Space assets have become critical to the well-being of humanity. There is a heavy reliance of modern societies on space vehicles and their applications. Furthermore, the use of satellites in the military field has widely proliferated.[182]

The integration of space-based assets into ground, air, and sea warfare has thus made them important not only for offensive purposes but also, if not especially, for defensive goals. Because of their relevance from economic and military perspectives, the protection of space assets constitutes a priority objective of the national security and defense strategies of the technologically advanced nations.

Consequently, the concept of space security has continually gained international relevance. A reminder of the dangerous balance existing in space and of the vulnerability of space objects occurred in January 2007, when China performed an anti-satellite test (ASAT). As discussed earlier, during the test China destroyed one of its aging satellites by means of a ground-based missile. The test had a deleterious effect on the space environment because thousands of pieces of debris were released upon impact. The United States responded to the Chinese test by destroying one of its malfunctioning satellites with a missile launched from a U. S. warship in 2008.

A number of initiatives aimed at enhancing the security of space objects and avoiding a military escalation in space have been put forward. Particular attention has been paid to the need to prevent the weaponization of outer space.

Regulating Military Activities in Outer Space

A coherent and unitary legal framework regulating military activities in outer space does not exist. The existing laws and regulations relevant to the military uses of outer space can be found in different agreements negotiated at different levels.

The fundamental framework is provided by the U. N. Charter, which aims at preserving international peace and security by obliging member states to refrain from the threat or use of force against the territorial integrity or political independence of any state.[183] The only two exceptions to the prohibition on the use of force are the right to act in sef-defense (Article 51, U. N. Charter) and the use of force authorized by the Security Council of the United Nations (Chapter VII of the Charter). Furthermore, the U. N. Charter requires U. N. members to settle their international disputes by peaceful means.

The U. N. Charter principles are relevant to outer space in virtue of Article III of the Outer Space Treaty, which establishes that space activities shall be carried on in accordance with international law, including the U. N. Charter. When applied to outer space, the U. N. Charter's principles mean that states shall refrain from their military threatening or attacking other countries space objects or nationals in space. At the same time, states are entitled to use the military to protect their space assets for reasons of self-defense or if authorized by the Security Council.

Apart from the Charter, the 1967 Outer Space Treaty is largely the only U. N. space treaty that includes provisions dealing with the military uses of outer space. It provides for the partial demilitarization of outer space, as it prohibits the placement of nuclear weapons and weapons of mass destruction in outer space or on celestial bodies and declares that celestial bodies shall be used exclusively for purposes. Nevertheless, the Outer Space Treaty does not prohibit other military uses of outer space, such as the deployment of military satellites and conventional weapons in outer space, the testing of weapons other than nuclear weapons and weapons of mass destruction, and the transit of intercontinental ballistic missiles. It is normally understood that the use of outer space for military purposes is legal as long as it is consistent with the U. N. Charter.

Because the Outer Space Treaty leaves several issues unaddressed, the legal framework applicable to the military uses of outer space needs to be completed with rules and provisions to be found in arms control treaties. These treaties, which might have a multilateral or bilateral dimension, impose restrictions on the testing and use of nuclear devices (i.e., the 1963 Partial Test Ban Treaty and the Comprehensive Test Ban Treaty), on the number of intercontinental ballistic missiles (i.e., the Strategic Arms Limitation Treaties, Salt-I and II), and on the use of military techniques causing long-lasting effects on the environment, including outer space (i.e., the Convention on the Prohibition of Military or Any Other Hostile Use of Environmental Modification Techniques, ENMOD Convention).

Prevention of Weaponization of Outer Space

The topic of the prevention of weaponization of outer space has been extensively discussed in recent years. The term weaponization of outer space refers to the deployment of weapons of an offensive nature in space or on the ground with their intended target located in space. Due to the increasing importance of space assets and the consequent need to protect them, there is a widespread concern that states might eventually weaponize outer space. Considering that the space treaties do not impose any substantial limit to such a weaponization, initiatives aimed at creating legal barriers to such an option have been launched. In this respect, proposals having both binding[184] and non-binding character have been submitted.[185] However, none of them has managed to achieve global acceptance. For the purpose of our discussion, it is important to point out that discussions on the theme of the prevention of weaponization of outer space have taken place within the Conference on Disarmament, a forum established in 1979 to deal with multilateral disarmament issues.

The Commercialization of Outer Space

Defining 'commercial' space activities is somewhat controversial task. A commercial space activity can be defined as one in which a private entity puts its own capital at risk and provides goods or services mostly to other private subjects or consumers rather than to the government. Examples of these activities are direct-to-home satellite television (e.g., DirecTV and DishTV), and commercial fixed satellites that transmit voice, data and Internet services (such as Intelsat Ltd., SES Global, Luxembourg).

Alternatively, a broader understanding of a commercial space activity includes sales of consumer equipment by companies even though the satellite system is owned by the government. The chief example of this is the Global Positioning System (GPS) navigation satellite system that is owned by the U. S. Department of Defense (DOD), but has a vast array of consumer uses ranging from automobile navigation systems to cell phones to precision farming. The devices used by consumers around the world in their cars, on their boats, or carried on their persons are sold by commercial companies, but the satellite signal that makes them work is provided for free by DOD.

In European legal circles, the term 'commercial' refers to an activity undertaken with the goal of obtaining a profit. Instead, in the United States, the word 'commercial' is used with reference to an activity in which a private entity is involved. The 2010 National Space Policy of the United States defines commercial space activities. The term commercial, for the purposes of this policy, refers to space goods, services, or activities provided by private sector enterprises that bear a reasonable portion of the investment risk and responsibility for the activity, operate in accordance with typical market-based incentives for controlling cost and optimizing return on investment, and have the legal capacity to offer these goods or services to existing or potential nongovernmental customers.

Regardless of any specific definition, nowadays commercial space activities are characterized by two main features: (1) space services offered to governmental and non-governmental subjects in return for a certain price; and (2) private entities operating in the space market not only as clients but also as manufacturers of key space services. In this sense, private operators are progressively becoming competitors of governments in sectors that were previously under the exclusive control of states. This trend is particularly evident in the area of launching services. Another area of significant interest is that of suborbital tourism, as the private sector is running, almost exclusively, as a business.

In the following sections the areas of launching services and suborbital tourism will be given special attention. Due to the fact that the United States has the largest number of private entities involved both business as well as the most developed legislation regulating private space activities, special attention will be dedicated to the U. S. situation.

Launch Services

With the ever increasing application of space technologies for both military and civil purposes coupled with the growing number of space actors, the question of having reliable and easily accessible services to launch and deploy objects in outer space has become crucially important. Nowadays not only satellites, which are often manufactured and owned privately, have to be taken into orbit, but also technical equipment, scientific instruments, replacement materials and food for astronauts are needed in space. If one takes the example of the International Space Station, its crew depends on regular supplies from Earth. Moreover, with the continuing commercialization of space activities, the number of private payloads waiting to be launched is likely to increase.

Under these premises it is not surprising that the launching business has long turned commercial, although initially private operators continued to rely exclusively on governmental space agencies, and still today they have to depend on them to deploy satellites. Transportation fees for private payloads are high, and capacities are limited due to the restricted number of launchers and spaceports. Nevertheless, in recent years established launching services have started competing for customers mostly because of cuts in public space budgets.

The first commercial space transportation company, namely Arianespace, was established in 1980 by several space related entities from ten European states upon the initiative of the ESA.[186] Around 60 % of Arianespace's share is held by the French Centre National D'Etudes Spatiales (CNES) and two companies of Astrium.[187] Despite this fact, ESA is directly involved in the activities of Arianespace, not only because it develops Arianespace launchers but also because it financially supports the company and has taken over responsibility from CNES. Arianespace is probably the biggest actor in this field of commercial launching services. Up to now, it has commercialized its launch vehicle Ariane 5 as well as the medium Soyuz launcher (Fig. 6.1).

After the fall of communism and the breakup of the Soviet Union, the Russian Space Agency ROSCOSMOS set up a thriving cargo business. Russian launch providers include the International Launch Services (ILS), the International Space Company (ISC), Kosmotras (a joint project among Russia, Ukraine and Kazakhstan) and Eurockot Launch Services (a German/Russian company).

Three companies compete for commercial launches in the United States: Boeing Launch Services, Lockheed Martin Commercial Launch Services and Space Exploration Technology (Space X). Also the multinational company Sea Launch/Land Launch participates in this business. Recently China, with the Long March series launch vehicles,[188] and India, with the Polar Satellite Launch Vehicle (PSLV) and Geosynchronous Satellite Launch Vehicle (GSLV),[189] entered the launching business offering even lower prices.

Due to these developments the other governmental launching services had to take similar steps in order to attract clients and stay in business. However, the analysis of the current situation reveals that it in the long run the existing governmental launching services will not be able to satisfy the growing demand for space transport. This opens the way for the development of less expensive private launchers.

The first privately developed launch vehicles made their appearance only in the early 1990s boosted by the adoption of two important bills in the United States, the 1984 Commercial Space Launch Act and the 1990 Launch Services Purchase Act.[190] The first company to enter the launch market was Orbital Sciences that, in the late 1980s, started manufacturing its rockets, Pegasus and Taurus, with the goal

Fig. 6.1 ESA' launch vehicle, Ariane 5. (Courtesy of ESA at http://spaceinimages.esa.int/Images/2011/05/Ariane_5_flight_VA2023)

of carrying small satellites into low Earth orbit. These two rockets were successfully launched respectively in 1990 and 1994 and, since then, they have regularly brought payloads into Earth orbit. In January 2000, Orbital Science[191] began operating its low cost rocket Minotaur that has successfully carried out eight missions. Another private company involved in the launching business is Space X.[192] Space X's Falcon vehicle has carried into orbit payloads of private companies such as Bigelow Aerospace as well as of governments.

The development of privately built launchers in the United States received a significant boost in the 2010 National Space Policy. Indeed, such a policy adopts an approach strongly in favor of the private sector.[193] Significantly, two types of private launchers are currently governmentally supported: (a) commercial crew launchers; and (b) commercial cargo launchers.

Commercial Crew: On February 1, 2010, President Obama proposed a drastic change to the U. S. human spaceflight program in his FY2011 budget request to Congress. He suggested relying on the commercial sector instead of NASA to build and operate systems to transport people to and from low Earth orbit (LEO). This also includes taking astronauts back and forth from the International Space Station (ISS). He demanded that $6 billion over a period of 5 years (FY2011-2015) in NASA's budget to used to subsidize companies to develop "commercial crew" launch vehicles and spacecraft for LEO missions. He also requested the cancellation of the Constellation program, begun under President George W. Bush, under which NASA was to build new launch vehicles (Ares I and V) and a spacecraft (Orion) to take astronauts into LEO as well as on longer missions.

The proposal met with a high level of controversy and was vigorously debated in Congress. The 2010 NASA Authorization Act (P.L. 111–267), signed into law in October 2010, was a compromise wherein NASA was directed to develop its own crew space transportation system—the Space Transportation System (STS) and a Multi-Purpose Crew Vehicle (MPCV)—as well as fund the commercial crew concept, but at a lower funding level. According to the law, the STS/MPCV systems are supposed to work as backup for a commercial crew in case they do not materialize or fail.

President Obama's FY2012 budget request for NASA, issued in February 2011, was equally controversial because the congressional committees that oversee NASA argued that it contravened the compromise reached in the 2010 NASA Authorization Act. In particular, NASA requested more money than was authorized for commercial crew and less money than was authorized for the STS/MPCV.

Meanwhile, with the conclusion of the space shuttle program in 2011, NASA no longer disposes of a launch vehicle to send astronauts to the ISS. How long it will be needed for the development of commercial crew services is unclear, but it is likely that such a development will require several years. For the time being, NASA acquires crew transportation services from Russia at a cost of $450 million per year.

The tension between Congress and the White House over the commercial crew initiative continues, although with the success of SpaceX's commercial cargo missions (see below), the situation is improving. Nevertheless, for FY2011 and FY2012, Congress provided sharply less funding than the Administration requested, and this trend is maintained in FY2013. The request for FY2013 was $830 million, but the House approved only $500 million, while the Senate only $525 million.

NASA initially awarded contracts to five companies for Crew Transportation Concepts and Technology Demonstration, or CCDEV (commercial crew development) in February 2010: Blue Origin, Boeing, Paragon Space Development Corp., Sierra Nevada Corp. and United Launch Alliance. Another round of winners of the CCDEV2 competition were announced in April 2011: Blue Origin, Boeing, Sierra Nevada and SpaceX. Those contracts were awarded as Space Act Agreements (SAAs) where NASA can pay companies for meeting agreed-upon milestones but has little oversight or insight into what the companies are doing.

The CCDEV program has moved into what NASA calls the Commercial Crew Integrated Capability program for commercial companies to develop an integrated crew transportation system (spacecraft, launch vehicle, and ground systems). In August 2012, NASA selected "2 1/2" proposals; this means that it is fully funding two companies' proposals (SpaceX and Boeing) and partially financially supporting a third (Sierra Nevada). The SpaceX proposal is to use its Falcon rocket for commercial crews as it does for the commercial cargo program. Boeing and Sierra Nevada each are developing crew capsules only and plan to launch them using the Atlas V rocket built by the United Launch Alliance (ULA). ULA and NASA had an unfunded Space Act Agreement that allowed them to exchange information about how Atlas V can meet the commercial crew requirements. This agreement was concluded in October 2012.

Commercial Cargo: Before Obama launched the "commercial crew" initiative, NASA already had initiated a "commercial cargo" program to rely on the commercial sector to take cargo to ISS. This program, called COTS (Commercial Orbital Transportation Services), was needed because the Bush Administration had decided to conclude the space shuttle program once ISS construction was completed. The last space shuttle mission took place in July 2011. Two companies, SpaceX and Orbital Sciences Corp., were awarded Space Act Agreements to develop spacecraft to take cargo to the ISS beginning in 2011, but these dates were postponed to 2012.

SpaceX conducted the first launch of its Falcon 9 launch vehicle in 2009. A second launch with a test version of its Dragon spacecraft—which is designed to take first cargo, and, later, crew into orbit—was successful in December 2010. Two more SpaceX test flights were planned, but they were combined into a single mission that was launched on May 22, 2012. On that flight, Dragon met and berthed with the ISS on May 25. Dragon remained attached to the ISS's Harmony module until May 31, 2012, when it was successfully unberthed by the ISS crew, deorbited, and splashed down in the Pacific Ocean about 490 km southwest of Los Angeles. That ends the COTS program for SpaceX, which now transitions into the operational Commercial Resupply Service (CRS) phase. NASA contracted for 12 SpaceX CRS flights through 2015. The first, "SpaceX CRS-1," completed its mission successfully in October 2012 despite the failure of one the nine Falcon 9 engines and other anomalies. The other company involved in commercial cargo, namely Orbital Sciences Corp., has a $1.9 billion contract for at least eight resupply flights using its new Antares rocket and Cygnus spacecraft, which are also designed to be disposable (Fig. 6.2).

Suborbital Space Tourism

Affordable and safe commercial private human access to outer space has rapidly become one of the hottest topic of modern space law and has captured widespread popular imagination.[194] The public perception of commercial space travel has moved from being a mere fantasy to becoming a concrete possibility to be materialized in a short time. As a consequence, significant resources have been

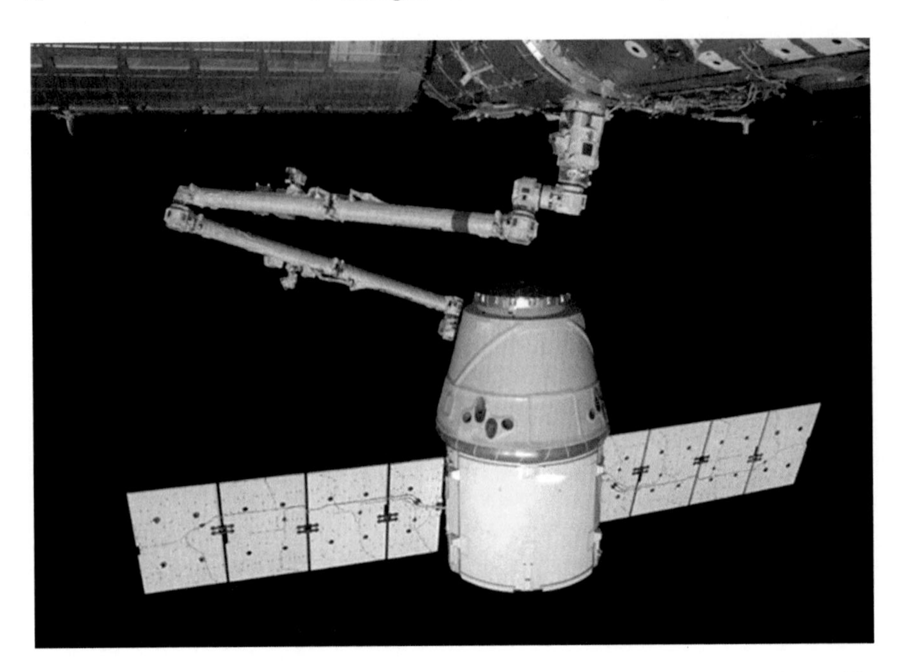

Fig. 6.2 SpaceX's Dragon capsule being 'captured' by the International Space Station on May 25, 2012. (Courtesy of NASA, at http://www.spacex.com/dragon.php)

invested towards developing Reusable Launch Vehicle (RLV) technologies, a key element in the formation of a space technology industry.

Within legal circles[195] 'space tourism' has been defined as "any commercial activity offering customers direct or indirect experience with space travel"[196] and a 'space tourist' as "someone who tours or travels into or through space or to a celestial body for pleasure and/or recreation."[197]

The era of space tourism began in 2001 when millionaire Dennis Tito became the first non-professional human being to reach outer space. Upon payment of an estimated $20 million fee, he was allowed to fly on board the Russian Soyuz capsule, to reach the International Space Station and to stay with professional astronauts onboard for some days. Tito's experience was repeated by other wealthy people in the years that followed.[198] This type of tourism is normally referred to as "orbital space tourism." In an orbital spaceflight orbital speed must be achieved to enable the vehicle to fly along the curvature of Earth and not fall back to Earth. An orbital spaceflight reaches a certain destination in LEO, in our case the ISS, where "tourists" spend a certain period of time (around 7–10 days). Normally, ISS tourists undergo a period of training before boarding the station.

Suborbital space tourism is different. Suborbital spaceflights do not reach orbital velocities. Usually, in a suborbital flight, a vehicle gets to an altitude of 100–200 km above sea level; after engine shutdown, a microgravity duration of 3–6 min is reached, at the conclusion of which the vehicle falls back to Earth. Suborbital tourists are not required to do any specific pre-flight training.

Fig. 6.3 A picture of Virgin Galactic's SpaceShipTwo and its carrier WhiteKnightTwo. (Courtesy of Virgin Galactic at http://www.virgingalactic.com/Used by permission)

Recently, the idea of suborbital space tourism has become quite popular. Interest in engaging in such tourism significantly grew in 2004 when Burt Rutan's Scaled Composites won the Ansari X-Prize award for using its SpaceShipOne to send a pilot over that threshold and back to Earth twice within 14 days (different pilots flew the craft each time). However, suborbital tourism is not a reality yet; indeed, so far no paying passenger has ever been on board a suborbital flight.

Nevertheless, the era of suborbital space tourism seems to be approaching fast. Richard Branson's company, Virgin Galactic,[199] has announced the intention to initiate suborbital flights services for paying customers in 2013 using its SpaceShipTwo spacecraft. Already for many years Virgin Galactic has been selling tickets for a seat on board one of its suborbital flights for a price of approximately $200,000. Thousands of people supposedly have bought such tickets. Other companies are also interested in the commercial suborbital market for experiments, people or both.

On August 9, 2011, NASA announced the selection of six companies for indefinite-delivery indefinite-quantity (IDIQ) contracts under its Commercial Reusable Suborbital Research (CRuSR) program to provide suborbital launch services for NASA technology experiments. The total value of all the contracts is $10 million. The companies selected are: Armadillo Aerospace, Near Space Corp., Up Aerospace Inc., Virgin Galactic, Whittinghill Aerospace LLC and XCOR (Fig. 6.3).

The Dual Nature of Satellites and the Commercialization of Outer Space

One of the inherent characteristics of satellites is to have a dual nature. This means that the same satellite might be used both for civilian (i.e., weather observation) and military purposes (intelligence from space). It is virtually impossible to prevent a satellite from performing both functions or to know, in real time, for which goal a satellite operator is using its space object.

The dual character of the majority of space objects may constitute an obstacle to the full development of the commercialization of outer space. Due to the uncertainty related to the actual use of a satellite and the security concerns that this fact generates, private operators in particular may face trouble in obtaining authorization to carry out space activities. Furthermore, substantive barriers can be imposed on the selling of privately owned or built satellites on the international market.

Chapter 7
The Future Challenges of Space Law and Policy

Space activities have undergone tremendous changes in the past decades. Once constituting a mere strategic tool in the hands of few states, now they are a global phenomenon affecting the lives of millions of people as well as international relations.

Firstly, the economic significance of space activities is rising. Revenues connected to the access and use of outer space are expanding, thus attracting numerous subjects towards space business. Secondly, the number of actors involved in space-related activities is growing. This includes not only countries belonging to the less-developed world, but also, and in particular, private operators. Finally, new developments have enabled space technologies to be applied for both military and civil purposes.

These factors have modified the nature of space activities and the goals attracting new participants to space-related businesses. The existing legal framework governing human operations in outer space was drafted in an era where outer space was perceived in a different manner and used for different reasons. Therefore, in order to enable twenty-first century space activities to remain orderly and peaceful as well as to fully fulfill their economic potential, some international regulatory and political changes are needed. Among the most significant challenges to be dealt with by the international community in the near future the following can be listed.

International Regulation of Space Debris

Earth's orbits are becoming progressively more crowded. Governmental and private entities are placing an increasing number of satellites attracted by the economic and social benefits deriving from their use. However, these kinds of activities are seriously undermined by the presence of orbital space debris. Space debris threatens the functionality of satellites and thus put at risk the investments made by space actors in the satellite business.

F. Tronchetti, *Fundamentals of Space Law and Policy*,
SpringerBriefs in Space Development, DOI: 10.1007/978-1-4614-7870-6_7,

As previously described no international binding rules regulating the prevention, disposal and removal of space debris exist. Based on the 2002 IADC Guidelines, several countries have adopted national legislation on space debris applicable to national space operators. Although these laws have contributed to improving the regulation of space debris and the behavior of space actors, they might not represent the optimal solution in the long run. Instead, the time might have come for states to agree on a set of binding technical and legal measures regulating the prevention and management of space debris during all phases of a space activity. Such measures might constitute the only solution to ensure that all space actors, whether they be governments or private entities, act in a consistent and predictable way with respect to the protection of the space environment and the safety of space objects.

Harmonization of National Space Legislation

One of the most important recent developments in the field of space law is the enactment of national space legislation. Even countries with very limited national space activities have drafted national space laws. One can observe that a small group of states, both space-faring and developing ones, are considering elaborating and enacting national space legislation in the near future. The reasons behind this choice include securing foreign investments, positioning themselves as attracting location for launching of space objects, and regulating national space activities.

Practice reveals that national space legislation tends to differ among each other due to the specific needs of each state and practical considerations. Such a diversity is not a positive development because it generates confusion and uncertainty concerning the law applicable to space activities. On one side, this may result in inconsistent behaviors by space actors authorized by different national authorities. On the other side, it may lead to cases of forum shopping, where private operators apply for a license to carry out space activities in those countries offering the most favorable legislative environment.

It is thus not surprising that on the international level initiatives have been undertaken to define a more harmonized approach towards national space legislation. For example, in 2008 a working group to address the topic "General information on national legislation relevant to the peaceful exploration and use of outer space" was established in the Legal Subcommittee of UNCOPUOS. The activities of this working group resulted in the adoption of a Draft Report containing elements that could be considered by countries when enacting regulatory frameworks for national space activities. The working group did not go as far as to suggest a model for national space legislation, nor did it reach agreement on practical measures to implement its suggestions. Nevertheless, in the coming years the international community should continue in its efforts aimed at ensuring a better harmonization of national space legislations; in this respect, the main obstacle will be to convince nations that following a harmonized approach does not mean they lose the right to independently choose the form and substance of their own national framework governing space activities.

Wider Ratification of the U. N. Space Treaties

Due to the increased dimension of space activities as well as the number of subjects involved in them, it is important that a common legal foundation be in place to properly regulate activities in space. Such a foundation is provided by the U. N. space treaties. Consequently, participation in and implementation of the space treaties should be encouraged. In this respect, the efforts undertaken by the Legal Subcommittee of UNCOPUOS to enhance adherence and compliance with the U. N. space treaties are a step in the right direction and should be continued in the years to come.

Regulation of Suborbital Space Tourism

Suborbital space tourism constitutes one of the most exciting perspectives in the area of space activities as well as a potentially profitable business. This type of tourism raises several legal questions that the international community needs to address in the near future. Firstly, the establishment of a boundary between airspace and outer space. As a sub-orbital journey involves a vehicle reaching low orbital altitudes and crossing both national airspace and outer space, agreeing on where airspace ends and outer space begins is crucial to clarify what type of law applies to different moments of such a journey. This issue is connected to another one, namely the applicability of existing aviation rules to sub-orbital flights crossing national airspaces. Finally, the legal status of private spaceflight participants should be defined, as it creates questions of liability and humanity. Significantly, initiatives addressing these issues have been launched both in academic circles and international organizations. Similar efforts should be broaden so as to elaborate an adequate legal framework for the era of sub-orbital flights.

Sustainability of Space Activities

The long-term sustainability of space activities is threatened by several factors, such as environmental concerns, orbital congestion, military activities, and in-orbit behaviors. Such a sustainability might be enhanced through the formulation of best practices, standards, and rules applicable to space operators. The international community should undertake efforts to develop such rules both in the form of non-binding norms or, preferably, of binding obligations.

In conclusion, no one can deny that the fulfillment of the above issues will be a challenging task. Nevertheless, a glimpse of hope is given by the fact these issues create real and tangible difficulties to the continuation and success of space activities. This fact could contribute to generate the political will and the diplomatic conditions enabling States to seriously deal with them.

Conclusion

Top 10 Things to Know About Space Law and Policy

1. Space Activities Are Regulated By a Mix of International and National Rules.

The legal foundation for space activities is provided by five international treaties and a set of principles adopted by the United Nations. These rules are complemented by an increasing number of space laws enacted at the national level. Other relevant provisions can be found in the context of space-related international organizations.

2. The United Nations Plays a Central Role in the Law-Making Process Related to Outer Space.

The main international rules applicable to human activities in outer space have been elaborated within the United Nations. A dedicated committee to deal with technical and legal issues connected to space activities exists, the United Nations Committee on the Peaceful Uses of Outer Space (UNCOPUOS). Although the law-making activity within UNCOPUOS has gradually slowed down, the committee remains the principal forum for discussion of space-related issues such as voluntary measures to decrease the creation of orbital debris and to pursue in a constructive way the sustainability of space.

3. The Adoption of Soft Law Instruments Currently Constitutes the Most Viable Method for Addressing Space Issues on the International Level.

For several years states have been unable to agree on new binding international norms on space matters. The emergence of key issues in the area of space activities has forced the international community to pursue regulatory measures to address them. The adoption of non-binding instruments, the so-called soft laws, has emerged as the most workable way to achieve this goal. Soft law documents have typically been formulated in the context of inter-governmental and non-governmental organizations.

F. Tronchetti, *Fundamentals of Space Law and Policy*,
SpringerBriefs in Space Development, DOI: 10.1007/978-1-4614-7870-6,

4. Space Actors Include States, Inter-governmental Organizations and Private Entities

During the first decades of the space era space activities were the exclusive domain of states. Increasingly, thanks to a favorable regulatory environment and attracted by potential profits, private entities have actively entered the space business. The impact of private operators in the space market is constantly growing. Additionally, some inter-governmental organizations play an important part in space matters.

5. Space Activities Must Be Carried Out in Conformity With Fundamental International Law Principles.

International space law is a branch of public international law. All activities in outer space must be undertaken in full compliance with fundamental principles of international law, particularly those included in the U. N. Charter. This idea is enshrined in Article III of the 1967 Outer Space Treaty. Although there are exceptions, most countries seek to conform to the provisions of the U. N. treaties and other relevant elements of space law and regulations.

6. Outer Space Is Free for Exploration and Use By All Nations Without Discrimination.

All states have the right to freely access, explore and use outer space. No country needs to obtain authorization to perform space activities or can be discriminated against based on its level of economic development. This principle is laid down in Article I of the 1967 Outer Space Treaty.

7. No Part of Outer Space, Including Celestial Bodies, Is Subject to National Appropriation.

This is a cardinal provision of space law that clearly helps to distinguish the legal situation on Earth from that in space. On Earth, states exercise their sovereign authority over physical territories. Appropriation by claim of sovereignty, use, and occupation were the traditional ways in which states would claim unowned areas. International space law prohibits such appropriation and establishes that outer space and its celestial bodies are international areas (sometimes referred to as a 'common'), which cannot be appropriated by anybody. Such a prohibition also extends to all private entities engaged in space activities. This provision makes private ownership claims over celestial bodies fallacious and void of any legal significance. The non-appropriation principle is provided for in Article II of the 1967 Outer Space Treaty.

8. The Commercialization of Space Activities Is a Growing Trend in the Twenty-First Century.

The increasing application of space technologies, coupled with the ever-increasing request for space services in the civilian context, represent driving forces for the commercialization of outer space. Sectors previously under strict governmental control have been made more accessible to the private sectors and open to commercial endeavors. This trend has been particularly visible in the field of manufacturing and operation of communications satellites and in the launching business, but is now evolving into areas such as commercial space travel and private space habitats.

9. The Long-Term Sustainability of Space Activities is Threatened by Environmental and Security Concerns.

The possibility to carry out activities in outer space in the years to come is undermined by environmental issues and security concerns. On one side, a vast number of orbital debris threatens the safety of space objects. On the other side, the defense implications of satellites make such space facilities a potential military and strategic target. The international community has undertaken soft law initiatives to address these issues. A more traditional approach, resulting in hard law binding obligation, might be preferable in the long run.

10. The Exploration and Use of Outer Space Influence International Relations and International and National Decision-Making.

Space activities have a profound impact on modern societies both from a civilian and military perspective. Millions of people benefit from space applications on a daily basis. Furthermore, space technologies significantly augment the efficiency and precision of military operations on the ground. Consequently, ensuring free access to and use of outer space and guaranteeing security of space objects are priorities of the political agenda of the most advanced nations. National space policies and strategies delineate a country's approach towards space-related issues on a national and global scale.

End Notes

1. This definition is taken from the Law.com dictionary, and it is accessible at http://dictionary.law.com/Default.aspx?selected=1111.
2. For a definition of 'space law' see F. Lyall/P. Larsen, Space law: a treatise, Ashgate Publishing (2009), p. 2; I.H.Ph. Diederiks-Verschoor/V. Kopal, An introduction to space law, Kluwer Law International (2008), p. 6.
3. V. Kopal, Outer space—A legal issue, in C. Brünner, A. Soucek, Outer space in society, politics and law, Springer Wien, New York (2011), p. 224;
4. The 100 km demarcation has also been accepted by the Fédération Aéeronatique Internationale (FAI). The FAI promulgates standards and keeps record of space activities. However, the FAI is a non-governmental body; thus its views are not binding upon States. Significantly, also the Australian Space Law Act sets the demarcation between airspace and outer space at an altitude of 100 km from sea level.
5. The 100 km margin is based on the Karman line theory. According to this theory, which is also known as the "aerodynamic lift theory", everything beyond 100 km above sea level is considered as 'outer space' because starting from this altitude any aircraft would have to fly faster than orbital speed to get enough aerodynamic lift to support itself.
6. See http://www.thefreedictionary.com/policy.
7. See F. Lyall/P. Larsen, supra footnote 2, pp. 31 ss.
8. G. Lafferranderie, Introduction, in G. Lafferranderie (ed.), Outlook on space law over the next 30 years, Essay published for the 30th anniversari of the Outer Space Treaty, Kluwer Law International (1997), pp. 2–5.
9. V. Mandl, Das Weltraum-Recht. Ein Problem der Raumfahrt. Mannheim, Berlin, Leipzig: J. Bensheimer (1932).
10. For a description of the early years of space law see V. Kopal, Evolution of the doctrine of space law, in N. Jasentuliyana (ed.), Space law, development and scope, Westport, Connecticut, London, 1992, pp. 17–32.
11. UNGA Resolution 1348 (XIII), "Questions of the Peaceful Use of Outer Space", 13 December 1958.

F. Tronchetti, *Fundamentals of Space Law and Policy*,
SpringerBriefs in Space Development, DOI: 10.1007/978-1-4614-7870-6,

12. UNGA Resolution 1472 (XIV), "International Cooperation in the Peaceful Uses of Outer Space, 12 December 1959.
13. Generally, on this point see V. Kopal, United Nations and the progressive development of international space law, 7 Finnish Yearbook of International Law 1(1996).
14. UNGA Resolution 1721 (XVI), 20 December 1961.
15. UNGA Resolution 1962, 13 December 1963.
16. On the topic of soft law see A. Boyle, Soft law in international law-making", in M.D. Evans. (ed.), International law, (Oxford University Press, 2nd edition (2006), pp. 141–158; Andrew T. Guzman, Timothy L. Meyer, Explaining soft law, Berkley Program in Law and Economics, WP series, (2009).
17. For a description of the role of soft law in outer space see F. Tronchetti, Soft law, in n C. Brünner, A. Soucek, Outer space in society, politics and law, Springer Wien, New York (2011), pp. 619–637.
18. For a general analysis of the five UN space treaties see C.Q. Christol, The modern international law of outer space, Pergamon Press (1982).
19. For a comprehensive description of the provisions of the Outer Space Treaty see S. Hobe/B.Schmidt-Tedd/K.U. Schrogl (eds.), Cologne Commentary on Space Law, Vol. I—The Outer Space Treaty, Carl Heymanns Verlag, (2009).
20. For example, this is the argument used by some private companies selling extraterrestrial properties on the web. For a detailed discussion of this topic see V. Pop, The men who sold the Moon: science fiction or legal nonsense?, 17 Space Policy 195, (2001); F. Tronchetti, The non-appropriation principle as a structural norm of international law: a new way of interpreting Article II of the Outer Space Treaty, 33 Air Space Law 277, issue 3, (2008).
21. For a detailed analysis of the Rescue and Return Agreement see F.G. von der Dunk, A sleeping beauty awakens: the 1968 Rescue Agreement after forty years, 34 Journal of Space Law (2008), pp. 411–434.
22. On the legal status of space tourists see F.G. von der Dunk, Space for tourism? Legal aspects of private spaceflight for tourist purposes, 49 IISL Proceedings 18 (2006).
23. S.Gorove, Legal problems of the rescue and return of astronauts, 3 Int. Lawyer (1968–1969), pp. 898–902; I.H. Ph. Diedericks-Verschoor, Search and rescue in space law, Proceedings of the Nineteenth Colloquium on the Law of Outer Space 17 (1977).
24. M.J. Sundahl, Rescuing space tourists: a humanitarian duty and business need, 50 IISL Proceedings 204 (2007).
25. An analysis of the provisions of the Liability Convention is provided in A. Kerrest, Liability for damage caused by space activities, in M. Benkö/ K.U. Schrogl (eds.), Space law: current problems and perspectives for future regulations, Eleven International Publishing (2005), pp. 91–112.
26. A damage is a "loss of life, personal injury or other impairment of health, or loss of or damage to property of States or of persons, natural or judicial, or property of international intergovernmental organizations", Article I, a, Liability Convention.

27. The State which launches an object into space, the so-called "launching State", can be:) a State which launches or procures the launching of a space object; 2) a State from which territory or facility a space object is launched. It should be noted that the word 'launching' includes attempted launching", Article I, b, Liability Convention.
28. For an analysis of the Claim Commission see later.
29. For a detailed analysis of the 1975 Registration Convention see B. Schmidt-Tedd/ M. Gerhard, Registration of space objects: which are the advantages for States resulting from registration, in M. Benkö/K.U. Schrogl (eds.), Space law: current problems and perspectives for future regulations, Eleven International Publishing (2005), pp. 121–140.
30. Article II, Registration Convention.
31. Article III, Registration Convention.
32. For a description of the Moon Agreement see F. Tronchetti, The exploitation of the natural resources of the Moon and other celestial bodies: a proposal for a legal regime, Martinus Nijhoff Publishers (2009); F.G. von der Dunk, The Moon Agreement and the prospect of commercial exploitation of lunar resources, 32 Annals Air & Space L., (2007).
33. On the concept of the 'common heritage of mankind see generally H.S. Rana, The Common Heritage of Mankind & the Final Frontier: A Revaluation of Values Constituting the International Legal Regime for Outer Space Activities, 26 Rutgers L.J. 225, (1994).
34. In the Moon Agreement any provision relating to the Moon shall be deemed to apply also to celestial bodies (Art. 1, para 3).
35. J.W. Benson, Space resources: first-come first-served, in Proceedings of the Forty-First Colloquium on the Law of Outer Space 46 (1998); R. Berkley, Space law versus space utilization: the inhibition of private industry in outer space, 15 Wisconsin Intern. L. Journ. 421, (1996–1997); R. Buxton, Property in outer space: the common heritage of mankind principle vs. the first in time, first in right rule of property law, in 69 J. Air L. & Com. 689 (2004).
36. V. Kopal, Outer space—A legal issue, in C. Brünner, A. Soucek, Outer space in society, politics and law, Springer Wien, New York (2011), p. 229.
37. Principles Governing the Use by States of Artificial Earth Satellites for International Direct Television Broadcasting, Dec. 10, 1982, UN Doc. AlRes/ 37/92. G.A. Res. 37/92. U.N. GAOR, 37th Sess .Supp. No. 51, at 98, U.N. Doc. Al37/51.
38. Principles Relating to Remote Sensing of the Earth from Outer Space, Dec. 3, 1986, U.N. GAOR, 41st Sess., Supp. No. 53, at 115, UN. Doc. Al41153, UNGA Res. 41/65 (1986).
39. Principles Relevant to the Use of Nuclear Power Sources in Outer Space, Dec. 14, 1992, U.N. Doc. A/Res/47/68.
40. Declaration on International Cooperation in the Exploration and Use of Outer Space for the Benefit and in the Interest of All States, Taking into Particular Account the Needs of Developing Countries, Dec. 13, 1996, U.N. Doc. AJRes/511122.

41. For a general analysis on the UNGA Principles see, pp. 349–83; V. Kopal, United Nations and the progressive development of International Space Law, 3 Finnish Yearbook of International Law 1 (1997).
42. For an analysis of the status of UNGA Resolutions see A. Terekhov, UN General Assembly Resolutions and Outer Space Law, Proceedings of the Fortieth Colloquium on the Law of Outer Space 87 (1997).
43. S. Marchisio, The 1986 United Nations Principles on Remote Sensing: A Critical Assessment, in Scritti in onore di Gaetano Arangio-Ruiz, pp. 1311–1340 (Naples, 2004).
44. See Principle XII, Remote Sensing Principles.
45. S. Langston, A Comparative Legal Analysis of US and EU Data Access Policies for Earth Remote Sensing, Contemporary Issues and Future Challenges in Air & Space Law 277 (2011).
46. For a description of the NPS principles see A. Soucek, International law, in C. Brünner, A. Soucek (eds.), Outer space in society, politics and law, Springer Wien, New York 2011, pp. 373 ss.
47. See Principle III, Nuclear Power Sources Principles.
48. For an analysis of the Space Benefits Declaration see M. Benkö/K.U. Schrogl, The 1996 UN-Declaration on space: ending the north-south debate on space cooperation, in Proceedings of the Thirty-Ninth Colloquium on the Law of Outer Space (1996), p. 183.
49. S. Hobe, Space Law—An analysis of its Development and its future, (C. Brünner, A.Soucek), Outer Space in Society, Politics and Law, Springer Press (2011), pp. 476–490.
50. UNGA Res. 62/101 (17 December 2007) 'Recommendations on Enhancing the Practice of States and Intergovernmental Organizations in Registering Space Objects', UN .Doc. A/Res/62/101.
51. Convention for the Establishment of a European Space Agency (hereafter ESA Convention), Paris, done 30 May 1975, entered into force 30 October 1980; UKTS 1981 No. 30; 14 ILM 864 (1975).
52. See Article XV, ESA Convention.
53. Article XIV(1), ESA Convention.
54. On the issue of space debris see K.U. Schrogl, Space and its sustainable uses, in C. Brünner, A. Soucek, Outer space in society, politics and law, Springer Wien, New York (2011), pp. 605–607; L. Viikari, The environmental element in space law, Martinus Nijhoff/Brill Publishers (2008), pp. 31 ss.
55. See for example, Technical Report on space debris, United Nations General Assembly. Technical report of the Scientific and Technical Subcommittee on space debris. UN Doc. A/AC.105/720, 1999.
56. See the Inter-Agency Space Debris Coordination Committee's website at http://www.iadc-online.org.
57. The Space Debris Mitigation Guidelines of UNCOPUOS has been published in 2010 by the United Nations Office for Outer Space Affairs (UNOOSA) and are available at http://www.unoosa.org/pdf/bst/COPUOS_SPACE_DEBRIS_MITIGATION_GUIDELINES.pdf.

58. For example, guidelines similar to the Governmental Orbital Debris Mitigation Standard Practices have been developed in the form of regulation by the US Federal Communications Commission, the Department of Transportation and the Department of Commerce. For more information see N.L. Johnson, Orbital debris research in the US, Proceedings of the Fourth European Conference on Space Debris, ESA/ESOC, Darmstadt/Germany 18–20 April 2005, pp. 5–10, 9.
59. Commercial Space Transportation Licensing Regulation, Part 415.39. For more information on this Regulation see L. Viikari, supra footnote 54, p. 108.
60. See Report of the Scientific and Technical Subcommittee on its 42nd session, Vienna, 21 February–4 March 2005, UN Doc. A/AC.105/786, para. 91; R. Tremayne-Smith, Environmental protection and space debris issues in the context of authorization, in F.G. von der Dunk (ed.), National space legislation in Europe, Issues of authorization of private space activities in the light of developments in European space cooperation, Martinus Nijhoff/Brill Publishers (2011), pp. 179–188.
61. The text of the EU Draft Code of Conduct for Outer Space Activities is available at: http://www.consilium.europa.eu/showPage.aspx?id=1570&lang=EN.
62. See Section 5 of the EU Draft Code of Conduct: Measures on space debris control and mitigation, and also Arts. 4.2 and 4.3.
63. See Joseph N. Pelton, Orbital Debris and Other Space Threats, Springer Press, New York (2013) for a description of the Space Data Association and its current membership.
64. See http://www.unidroit.org/dynasite.cfm?dsmid=103283.
65. See http://rescommunis.olemiss.edu/2012/03/12/unidroit-adopts-space-assets-protocol/; http://www.satellitetoday.com/via/globalreg/New-ITU-Role-in-the-UNIDROIT-Space-Asset-Protocol_39279.html.
66. C. Contant-Jorgenson, P. Lala, K.U. Schrogl (eds.), Cosmic study on space traffic management. IAA: Paris (2006).
67. This trend continues today. For example, the Obama Administration further cut funds for its civil space program in early 2010, see Associated Press, “Obama to cut NASA’s Moon plan: officials”, 31 Jan. 2010. CBC News 05 Oct. 2010.
68. In this respect see R. Jakhu (ed.), National regulations of space activities, Heidelberg-London-New York, Springer Press (2010); M. Gerhard, K.U. Schrogl, Report of the Project 2001, Working Group on national space legislation, in K.H. Böckstiegel (ed.), Project 2001—Legal framework for the commercial use of outer space, Cologne: Carl Heymans Verlag (2002), pp. 530 ss.
69. See I. Marboe, National space legislation, in C. Brünner, A. Soucek (eds.), Outer space in society, politics and law, Springer Wien, New York (2011), pp. 439–455.
70. For an analysis of Art. VI of the Outer Space Treaty see F.G. von der Dunk, Article VI of the Outer Space Treaty, S. Hobe, B. Schmidt-Tedd,

K.U. Schrogl (eds.), Cologne Commentary on Space Law, Vol. I, Cologne, Carl Heymanns Verlag (2009), pp. 117 ss.; F.G. von der Dunk, The origins of authorization: Article VI of the Outer Space Treaty and international law, in F.G. von der Dunk (ed.), National space legislation in Europe, Martinus Nijhoff Publishers/Brill, 2012, pp. 3 ss.

71. Cassese, International Law, Oxford University Press (2001), pp. 187–191.
72. J. Hermida, Legal basis for national space legislation, Dordrecht: Kluwer Academic Publisher (2004), pp. 29–32.
73. S. Hobe, Harmonization of national law as an answer to the phenomenon of globalization, in K.H. Böckstiegel (ed.) Project 2001—Legal framework for the commercial use of outer space, Cologne: Carl Heymans Verlag (2002), pp. 639–640.
74. See Marboe, supra footnote 69, p. 443.
75. Generally see J. Gabrynowicz, One half century and counting: the evolution of US national space law and three long-term emerging issues. 4 Harward law and policy review 405 (2010). See 1984 Commercial Space Launch Act, Public Law 98–575, 98th Congress, H.R. 3942, 10 December 1984.
76. 1984 Commercial Space Launch Act, 14 C.F.R., 440.9 (c).
77. 1984 Commercial Space Launch Act, 14 C.F.R., 440.9 (e).
78. 1984 Commercial Space Launch Act, 14 C.F.R., 440.19 (a).
79. LOI n. 2008-518 du 3 Juin 2008 relative aux operations spatiales.
80. Article 6, French Space Operation Act.
81. Article 13, French Space Operation Act.
82. Article 14, French Space Operation Act.
83. Article 15, French Space Operation Act.
84. France can be considered the third space-faring countries of the world and the first launcher in Europe.
85. P. Achilleas, Regulation of space activities in France, in R. Jakhu (ed.), National regulation of space activities, Dordrecht: Springer Press (2010), p. 111.
86. Loi relative aux opérations spatiales, LOI n. 2008-518 du 3 Juin 2008.
87. See supra at p. 29.
88. Article 13, Belgian Space Law Act.
89. Article 14, Belgian Space Law Act.
90. Articles 15–17, Belgian Space Law Act.
91. Royal Decree implementing certain provisions of the law of 17 September 2005 on the activities of launching flight operations and guidance of space objects.
92. Section 3(4), Dutch Space Law Act.
93. Ibidem.
94. Section 12 (1), Dutch Space Law Act.
95. Section 12 (2), Dutch Space Law Act.
96. For information on the US satellite export control policy see J. Hillery, "U.S. satellite export control policy", Center for Security and International Studies (Sept. 20 2006).

97. The United States Munitions List (USML) is a list of articles, services, and related technology designated as defense- and space-related by the United States federal government. This designation is pursuant to sections 38 and 47(7) of the Arms Export Control Act (22 U.S.C. 2778 and 2794(7)).
98. The US policy towards the export and import of commercial communication satellites and related components has changed several times. The issue has been whether these satellites should fall under the jurisdiction of the State Department or the Department of Commerce, the latter ensuring a more liberal and less-restrictive approach.
99. On this point see M. Mineiro, An inconvenient regulatory truth: Divergence in US and EU satellite export control policies on China, 27 Space Policy 213 (2011).
100. For an analysis of the European Policy on export control see F. von der Dunk, A European equivalent to the United States export controls: European Law on the control of international trade in dual-use space technologies, Astropolitics 7:2 (May 2009), pp. 101–134.
101. See Article 346, Consolidated version of the Treaty on the Functioning of the European Union.
102. Wetter, Enforcing European Union Law on Exports of Dual-Use Goods, Oxford University Press (2009), p. 49.
103. Article 4 (2), Council Regulation (EC) No 428/2009, Setting up a Community Regime for the Control of Exports, transfer, brokering, and transit of dual-use goods (re-cast), [2009] O.J. L 134. Council Regulation 428/2009 is the follow up to Council Regulation (EC) No. 1334/2000.
104. Article 3, Council Regulation (EC) No.428/2009.
105. Id. at Article 9 (2).
106. UNGA Resolution 1348 (XIII), "Questions of the Peaceful Use of Outer Space", 13 December 1958.
107. UNGA Resolution 1472 (XIV), "International Cooperation in the Peaceful Uses of Outer Space, 12 December 1959.
108. On the work and functioning of UNCOPUOS and its Legal Subcommittee see V. Kopal, The work of the Committee on the Peaceful Uses of Outer Space, in K.H. Böckstiegel (ed.), Project 2001, Legal framework for the commercial use of space, Cologne, Carl Heymanns Verlag (2002), pp. 17–26.
109. On the functioning of ITU see F. Lyall, International Communications: the International Telecommunication Union and the Universal Postal Union, Ashgate (2011).
110. See http://www.unidir.org/html/en/about.html.
111. See T. Hitchens, Saving space: threats proliferation and mitigation, study commissioned in 2009 by the International Commission on Nuclear Non-Proliferation and Disarmament, available at http://www.icnnd.org/research/Hitchens_Saving_Space.pdf.
112. Preamble, 1975 European Space Agency Convention.
113. See http://www.apsco.int/.
114. See http://www.aprsaf.org.

115. For information on the Sentinel Asia initiative see http://www.jaxa.jp/article/special/sentinel_asia/index_e.html.
116. This definition is taken from http://www.thefreedictionary.com/safety.
117. On the issue of 'space safety' see: J.N. Pelton and R. Jakhu, Space Safety Regulation and Standards, (2010) Amsterdam, Elsevier Press.
118. For more information on the IAASS see at http://www.iaass.org/home0.aspx.
119. See the IAASS's website at http://www.iaass.org/books0.aspx.
120. See: http://www.spacesafetyfoundation.org.
121. See http://www.iafastro.com/index.html?title=About_us.
122. See http://www.ila-hq.org.
123. G.M. Goh, Dispute Settlement in International Space Law, A Multi-Door Courthouse System, Martinus Nijhoff/Brlll (2008), p. 81.
124. See generally F. Pocar, An Introduction to the PCA's Optional Rules for Arbitration of Disputes Relating to Outer Space Activities, 28 Journal of Space Law 171 (2011).
125. C.J. Cheng, International Arbitration System as Mechanism for the Settlement of Disputes Arising in Relation to Space Commercialization, 5 Singapore J. Int'l & Comp. L. 167 (2001).
126. Article I (2), Outer Space Treaty.
127. Article II, Outer Space Treaty.
128. Art. III, Outer Space Treaty.
129. For a description of the main methods of international dispute settlement see J.G. Merrils, International Dispute Settlement, 3rd ed., p. 1 (1998).
130. M.N. Shaw, International Law, Cambridge University Press 5th ed. 2003, pp. 175, 222–246.
131. Article IX, Liability Convention.
132. Article XII, Liability Convention.
133. UN Doc. A/AC.105/C.2/L.74.
134. Article XV, Liability Convention.
135. Aer. XVIII, Liability Convention.
136. On 24 January 1978 the Soviet satellite Cosmos 954 crashed into Canadian territory. The satellite caused severe pollution to the area of impact because it was fueled with nuclear power sources. Initially, Canada claimed compensation for 6 million dollars but eventually the case was settled when the Soviet Union agreed to pay 3 million dollars to Canada. As described, although both States were parties to the Convention, the case was settled between them bilaterally, without making recourse to the dispute settlement mechanism available under the Convention.
137. K.H. Böckstiegel, Settlement of Disputes Regarding Space Activities, 21 Journal of Space Law 1 (1993).
138. Final Draft of the Revised Convention on the Settlement of Disputes related to Space Activities, ILA, Report of the 68th Conference, Taipei, Taiwan, Republic of China, (1998) 249–267.
139. Article 1 (1), 1998 ILA Draft Convention.
140. Section II, 1998 ILA Draft Convention.

141. Section III, 1998 ILA Draft Convention
142. Article 3, 1998 ILA Draft Convention.
143. Article 4, 1998 ILA Draft Convention.
144. Article 5, 1998 ILA Draft Convention.
145. Article 6 (1), 1998 ILA Draft Convention.
146. Article 6, 1998 ILA Draft Convention.
147. Article 6 (2), 1998 ILA Draft Convention.
148. For an introduction to the PCA Outer Space Optional Rules see S. Hobe, The Permanent Court of Arbitration Adopts Optional Rules for Arbitration of Disputes Relating to Outer Space Activities, 61 ZLW 61. 4 (Jg.1/2012).
149. For information about the Permanent Court of Arbitration see http://www.pca-cpa.org/showpage.asp?pag_id=363.
150. The text of the 2010 UNICTRAL Rules is available at http://www.uncitral.org/pdf/english/texts/arbitration/arb-rules-revised/pre-arb-rules-revised.pdf.
151. T.H. Webster, Handbook of UNICITRAL Arbitration, London Sweet & Maxwell (2010).
152. New York United Nations Convention on the Recognition and Enforcement of Foreign Arbitral Awards, art. V, June 7, 1959, 330 U.N.T.S. 38, 21 U.S.T. 2517, Art. V
153. The PCA has adopted the following procedural Rules for the settlement of disputes involving States, international organizations and private entities: the PCA Optional Rules of Procedure for Arbitrating Disputes between Two States (1992), the PCA Optional Rules for Arbitrating Disputes between Two Parties of Which Only One is a State (1993), the PCA Optional Rules for Arbitration between International Organizations and States (1996), and the PCA Optional Rules for Arbitration of Disputes between International Organizations and Private Parties (1996).
154. Article 1 (2), PCA Outer Space Optional Rules.
155. Article 10 (4), PCA Outer Space Optional Rules.
156. Article 28 (7), PCA Outer Space Optional Rules.
157. Articles 4 (5), 9 (3–4), PCA Outer Space Optional Rules.
158. Article 17 (6, 7, 8), PCA Outer Space Optional Rules.
159. See C. Al-Ekabi, Space Policies, Issues and Trend, ESPI Report 42, 2012.
160. On the economic significance of space activities see C. Venet, The economic dimension, in C. Brunner/A. Soucek (eds.), Outer Space in Society, Politics and Law (Springer Wien New York, 2011), pp. 55 ss.; OECD, The Space Economy at glance, Paris, 2007.
161. OECD. Space 2030. Exploring the Future of Space Applications. Paris, OECD 2004, 35.
162. S. Pagkratis, Space Policies, Issues and Trend, ESPI Report 2009.
163. C. Al-Ekabi, Space Policies, Issues and Trend, ESPI Report 42, 2012.
164. On the political dimension of space activities see M. Sheenan, The international politics of space, London/New York: Routledge (2007).
165. 1999 Vienna Space Millenium Declaration.

166. National Aeronautics and Space Act of 1958. Pub. L. No.85-568, 72 Stat. pp 426–438, 29 July 1958.
167. Presidential Directive NSC-37. National Space Policy. 11 May 1978.
168. National Security Decision Directive N. 42. National Space Policy. 4 July 1982.
169. White House National Science and Technology Council. National Space Policy. 19 Sept. 1996.
170. US National Space Policy. 31 Aug. 2006.
171. National Space Policy of the United States of America. 28 June 2010.
172. See for example Commission of the European Communities. Communication. The European Community and Space Challenges, Opportunity and New Actions. COM (92) 360 final of 23 Sept. 1992, Brussels. Commission of the European Communities, Communication. The European Union and Space: Fostering Applications, Markets and Industrial Competition. COM (96) 617 4 Dec. 1996. Brussels.
173. Council of the European Union. Resolution on Developing a Coherent European Space Strategy. Doc. 1999/C 375/01 of 2 Dec. 1999, Brussels.
174. Commission of the European Communities. Communication. Europe and Space: Turning to a New Chapter. COM (2000) 597 final of 27 Sept. 200. Brussels. EU.
175. Council of the European Union. Resolution on the European Space Policy. Doc. 10037/07 of 25 May 2007
176. Article 189, Treaty on the Functioning of the European Union. For information see F.G. von der Dunk, The EU space competence as per the Treaty of Lisbon: sea change or empty shell?, in IISL Proceedings 2012, pp. 382–392.
177. Supreme Soviet of the Russian Federation. Law of the Russian Federation N. 5663-1. About Space Activity. 20 Aug. 1993.
178. Federal Space Program of the Russian Federation for 2006–2015. Major provisions. Approved by Resolution N. 635 of the Government of the Russian Federation. 22 Oct. 2005.
179. For a description of the Japanese Basic Space Plan see S. Aoki, Current status and recent developments in Japan's National Space Law and its relevance to Pacific Rim space law and activities, 35 Journal of Space Law (Issue 2, 2009), p. 10.
180. Information Office of the State Council of the People's Republic of China, "China's Space Activities in 2011", see at http://www.globalsecurity.org/space/library/policy/int/china-space-activities_2011.htm.
181. N. Peter, The changing geopolitics of space activities, 22 Space Policy 100 (2006).
182. In particular satellites enable navigation, real-time weather data, instantaneous communications, gather intelligence, conduct reconnaissance and surveillance, warn of missile attacks, and allow precision attacks by missiles.
183. Article 2 (4), UN Charter.

184. In 2008 Russia and China put forward a Draft Treaty on the Prevention of the Placement of Weapons in Outer Space and of the Threat or Use of Force against Outer Space Objects (PPWT) in 2008. The Draft Treaty has received mixed responses and it is currently being reviewed by its drafters.
185. Among the proposals for non-binding rules it is worth mentioning the 2010 EU Draft Code of Conduct for Outer Space Activities (available at http://www.consilium.europa.eu/showPage.aspx?id=1570&lang=EN) and the Canadian Initiative for the formulation of Confidence-Building Measures (working paper entitled 'The Merits of Certain Draft transparency and Confidence Building Measures and Treaty Proposals for Space Security', CD/1865, 5 June 2009).
186. For a description of Arianespace's history see the Arianespace website at http://www.arianespace.com/about-us/milestone.asp
187. For a list of the participating countries, companies and their relative share see in the Arianespace website at http://www.arianespace.com/about-us-corporate-information/shareholders.asp.
188. See at http://www.cgwic.com/.
189. See at http://www.isro.org/Launchvehicles/launchvehicles.aspx.
190. 42 U.S.C. 2465 d.
191. For information see the Orbital Science's website at http://www.orbital.com.
192. For information see the SpaceX website at http://www.spacex.com.
193. See Principle 2; Goal 1; Sector Guidelines: Commercial Space Guidelines, 2010 National Space Policy of the United States of America.
194. Generally on 'space tourism' see S. Freeland, Space tourism and the international law of outer space, in S. Baht (ed.) Space law in the era of commercialization, Easter Book Company (2010), pp. 16 ss.
195. For an analysis of the legal issues surrounding private spaceflight see F.G. von der Dunk, Space tourism, private spaceflights and the law: key aspects, 27 Space Policy (2011), pp. 146–152.
196. S. Hobe, J. Cloppenburg, Towards a new aerospace convention? Selected legal issues of space tourism, Proceedings of the 47th Colloquium on the Law of Outer Space.
197. Z.N. O'Brien, Liability for injury, loss or damage to the space tourist, Proceedings of the 47th Colloquium on the Law of Outer Space.
198. Mark Shuttleworth (2002); Gregory Olsen (2005); Anousheh Ansari (2006); Charles Simonyi (2007 and 2009); Richard Garriot (2008); Guy Laliberté (2009).
199. On Virgin Galactic's activities see http://www.spaceshiptwo.net/.

Appendix

Selected Reading

Books

M. Benkö/K.U. Schrogl (eds.), *Space law: current problems and perspectives for future regulations*, Eleven International Publishing (2005);
K.H. Böckstiegel (ed.), *Project 2001, Legal framework for the commercial use of space*, Carl Heymanns Verlag (2002);
C. Brünner, A. Soucek (ed.), *Outer space in society, politics and law*, Springer Wien, New York (2011);
A. Cassese, *International Law*, Oxford University Press (2001);
C.Q. Christol, *The modern international law of outer space*, Pergamon Press (1982);
I.H.Ph. Diederiks-Verschoor/V. Kopal, *An introduction to space law*, Kluwer Law International (2008);
G.M. Goh, *Dispute Settlement in International Space Law, A Multi-Door Courthouse System*, Martinus Nijhoff/Brlll (2008);
S. Hobe, B. Schmidt-Tedd, K.U. Schrogl (eds.), *Cologne Commentary on Space Law, Vol. I*, Carl Heymanns Verlag (2009);
S. Hobe, B. Schmidt-Tedd, K.U. Schrogl (eds.), *Cologne Commentary on Space Law, Vol. II*, Carl Heymanns Verlag (2013);
R. Jakhu (ed.), *National regulations of space activities*, Springer Press (2010);
N. Jasentuliyana (ed.), *Space law, development and scope*, Westport, Connecticut, London (1992)
G. Lafferranderie (ed.), *Outlook on space law over the next 30 years, Essay published for the 30th anniversary of the Outer Space Treaty*, Kluwer Law International (1997);
F. Lyall/P. Larsen, *Space law: a treatise*, Ashgate Publishing (2009);
J. G. Merrils, *International Dispute Settlement*, Cambridge University Press 3rd ed. (1998);
J. N. Pelton and R. Jakhu, *Space Safety Regulation and Standards,* Elsevier Press (2010);

F. Tronchetti, *Fundamentals of Space Law and Policy*,
SpringerBriefs in Space Development, DOI: 10.1007/978-1-4614-7870-6,

J. N. Pelton, *Orbital Debris and Other Space Threats*, Springer Press (2013);
M. N. Shaw, *International Law*, Cambridge University Press 5th ed. (2003);
M. Sheenan, *The international politics of space*, Routledge (2007);
F. Tronchetti, *The exploitation of the natural resources of the Moon and other celestial bodies: a proposal for a legal regime*, Martinus Nijhoff Publishers (2009);
L. Viikari, *The environmental element in space law*, Martinus Nijhoff/Brill Publishers (2008);
F.G. von der Dunk (ed.), *National space legislation in Europe: issues of authorization of private space activities in the light of developments in European space cooperation*, Martinus Nijhoff/Brill Publishers (2011);

Articles

C. Al-Ekabi, Space Policies, Issues and Trend, ESPI Report 42 (2012);
S. Aoki, Current status and recent developments in Japan's National Space Law and its relevance to Pacific Rim space law and activities, 35 Journal of Space Law 10 (2009);
[199]C.J. Cheng, International Arbitration System as Mechanism for the Settlement of Disputes Arising in Relation to Space Commercialization, 5 Singapore J. Int'l & Comp. L. 167 (2001);
L.M. Fountain, Creating Momentum in Space: Ending the paralysis Produced by the Common Heritage of Mankind Doctrine, in 35 Conn.L.Rev. 1753, (2003);
J. Gabrynowicz, One half century and counting: the evolution of US national space law and three long-term emerging issues. 4 Harward law and policy review 405 (2010);
J. Hillery, "U.S. satellite export control policy", Center for Security and International Studies (Sept. 20 2006);
S. Hobe, Common heritage of mankind- an outdated concept in international space law, in Proceedings of the Forty-First Colloquium on the Law of Outer Space 271(1998);
S. Hobe, The Permanent Court of Arbitration Adopts Optional Rules for Arbitration of Disputes Relating to Outer Space Activities, ZLW 61. Jg.1/2012, pp. 4–25;
V. Kopal, United Nations and the progressive development of International Space Law, 3 Finnish Yearbook of International Law 1 (1997);
S. Langston, A Comparative Legal Analysis of US and EU Data Access Policies for Earth Remote Sensing, Contemporary Issues and Future Challenges in Air & Space Law 271 (2011);
F. Lyall, Re-thinking ITU, in Proceedings of the Forty-Third Colloquium on the Law of Outer Space, 309 (2000);
S. Marchisio, The 1986 United Nations Principles on Remote Sensing: A Critical Assessment, in Scritti in onore di Gaetano Arangio-Ruiz 1311 (2004);
M. Mineiro, An inconvenient regulatory truth: Divergence in US and EU satellite export control policies on China, 27 Space Policy 2011, pp. 213–221;
N. Peter, The changing geopolitics of space activities, 22 Space Policy 100 (2006);

F. Pocar, An Introduction to the PCA's Optional Rules for Arbitration of Disputes Relating to Outer Space Activities, 28 Journal of Space Law 171 (2011).
V. Pop, Appropriation in outer space: the relationship between land and ownership and sovereignty on the celestial bodies, 16 Space Policy, 275 (2000);
V. Pop, The men who sold the Moon: science fiction or legal nonsense? 17 Space Policy 195, (2001);
H.S. Rana, The Common Heritage of Mankind & Th e Final Frontier: A Revaluation of Values Constituting the International Legal Regime for Outer Space Activities 26 Rutgers L.J. 225 (1994);
K.U. Schrogl, The UN General Assembly Resolution "Application of the concept of launching state, in Proceedings of the Forty-Eighth Colloquium on the Law of Outer Space 308 (2005);
M.N. SchmitT, International Law and Military Operations in Space, 10 MAX PLANCK YEARBOOK OF UNITED NATIONS LAW 89–125 (2006);
M.J. Sundahl, Rescuing space tourists: a humanitarian duty and business need, 50 IISL Proceedings 204 (2007);
L.I. Tennen, Article II of the Outer Space Treaty, the status of the Moon and resulting issues, in Proceedings of the Forty-Seventh Colloquium on the Law of Outer Space 520 (2004);
A.D. Terekhov, UN General Assembly Resolutions and Outer Space Law, in Proceedings of the Fortieth Colloquium on the Law of Outer Space 14 (1997);
F. Tronchetti, The non-appropriation principle as a structural norm of international law: a new way of interpreting Article II of the Outer Space Treaty, 33 Air Space Law 277, issue 3, (2008);
F. Tronchetti, Preventing the weaponization of outer space: is a Chinese-Russian-European common approach possible? 27 Space Policy 81 (2011);
F. Tronchetti, Soft law, in C. Brünner, A. Soucek (ed.), *Outer space in society, politics and law*, Springer Wien, New York 619 (2011);
F.G. von der Dunk, A sleeping beauty awakens: the 1968 Rescue Agreement after forty years, 34 Journal of Space Law (2008), pp. 411–434;
F.G. von der Dunk, Space for tourism? Legal aspects of private spaceflight for tourist purposes, 49 IISL Proceedings (2006), pp. 18–28;
F.G. von der Dunk, Space tourism, private spaceflights and the law: key aspects, 27 Space Policy 146 (2011);
F.G. von der Dunk, The Moon Agreement and the prospect of commercial exploitation of lunar resources, 32 Annals Air & Space L., (2007);

Treaties and Conventions

Charter of the United Nations (1945), 59 Stat. 1031, UNTS No. 993;
Treaty Banning Nuclear Weapon Tests in the Atmosphere, in Outer Space and Under Water, (1963), 480 UNTS 43;
Treaty on Principles Governing the Activities of States in the Exploration and Use of Outer Space, including the Moon and Other Celestial Bodies, London/Moscow/ Washington, (1967), 610 UNTS;

Agreement on the Rescue of Astronauts, the Return of Astronauts and the Return of objects Launched into Outer Space, (1968), 69 UNTS 119, SopS 45–46/1970;
Vienna Convention on the Law of the Treaties, (1969), 8 ILM 679, (1969), SopS 32–33/1980;
Convention for the Establishment of a European Space Agency, with Annexes, (1975), 14 ILM 864;
Convention on Registration of Objects Launched into Outer Space, (1975), 1203 UNTS 15;
Agreement Governing the Activities of States on the Moon and Other Celestial Bodies, (1979), 1363 UNTS 3;
2012 UNIDROIT Draft Protocol on Matters Specific to Space Assets;
2008 Draft Treaty on the Prevention of the Placement of Weapons in Outer Space and of the Threat or Use of Force against Outer Space Objects;
1977 Convention on the Prohibition of Military or Any Other Hostile Use of Environmental Modification Techniques (ENMOD Convention);

Declarations and Other Relevant Documents

U. N. Res. 1721 (XVI), (1961);
Declaration of Legal Principles Governing the activities of States in the Exploration and Use of Outer Space, UNGA Res. 1962 (XIII), 13 December 1963;
Principles Governing the Use of States of Artificial Earth Satellites for International Direct Television Broadcasting, UNGA Res. 37/92, 10 December 1982;
Principles Relating to Remote Sensing of the Earth from Outer Space, UNGA Res. 41/65, 3 December 1986;
Principles Relevant to the Use of Nuclear Power Sources in Outer Space, UNGA Res. 47/68, 14 December 1992;
Declaration on International Cooperation in the Exploration and Utilization of Outer Space for the Benefit and in the Interests of all States, Taking into Particular Account the Needs of Developing Countries, UNGA Resolution 51/122, 13 December 1996;
Resolution 59/115, 10 December 2004, Application of the concept of the "launching State";
Resolution 62/101, 17 December 2007, Recommendations on enhancing the practice of States and international intergovernmental organizations in registering space objects;
Inter-Agency Debris Committee Space Debris Mitigation Guidelines (2002, rev. 2007);
Space Debris Mitigation Guidelines of the Committee on the Peaceful Uses of Outer Space, Vienna, United Nations (2010);
European Union Draft Code of Conduct for Outer Space Activities (2010);
Supreme Soviet of the Russian Federation. Law of the Russian Federation N. 5663-1. About Space Activity. 20 Aug. 1993;

National Space Policy of the United States of America (2010);
Loi relative aux opérations spatiales, LOI n. 2008-518 du 3 Juin 2008;

Websites

Arianspace: http://www.arianespace.com/about-us/milestone.asp
Asia-Pacific Regional Space Agency Forum (APRSAF): http://www.aprsaf.org
Asia Pacific Space Cooperation Organization (APSCO): http://www.apsco.int/
China Great Wall Corporation: http://www.cgwic.com/
European Space Agency (ESA): http://www.esa.int
European Union (EU): http://europa.eu/index_en.htm
Indian Space Research Organization: http://www.isro.org/Launchvehicles/launch vehicles.aspx
Inter-Agency Space Debris Coordination Committee (IADC): http://www.iadc-online.org
International Association for the Advancement of Space Safety (IAASS): http://www.iaass.org/home0.aspx
International Astronautical Federation (IAF): http://www.iafastro.com/index.html?title=About_us
International Law Association (ILA): http://www.ila-hq.org
International Law Association (ILA): http://www.ila-hq.org
International Space University (ISU): http://www.isunet.edu/
International Telecommunication Union: http://www.itu.int
National Astronautics and Space Administration (NASA): http://www.nasa.gov
Orbital science: http://www.orbital.com
Permanent Court of Arbitration: http://www.pca-cpa.org/showpage.asp?pag_id=363.
Sentinel Asia initiative: http://www.jaxa.jp/article/special/sentinel_asia/index_e.html
SES Global: http://www.ses.com/4232583/en
Space foundation: http://www.spacesafetyfoundation.org
SpaceX http://www.spacex.com
UNIDROIT: http://www.unidroit.org/dynasite.cfm?dsmid=103283
United Nations Office for Outer Space Affairs (UNOOSA): http://www.oosa.unvienna.org
Virgin Galactic: http://virgingalactic.com

About the Author

Dr. Fabio Tronchetti has been an Associate Professor at the School of Law of the Harbin Institute of Technology, People's Republic of China, since September 2009. Before, he was Lecturer and Academic Coordinator at the International Institute for Air and Space Law, Leiden University, the Netherlands. He holds a PhD in International Space Law (Leiden University) and an Advanced LL.M (Master's of Law) in International Relations (Bologna University, Italy). He has participated as a speaker in several international space law conferences and has published several articles in the fields of space law and policy and European Law. He is the author of the book *The Exploitation of Natural Resources of the Moon and Other Celestial Bodies—A Proposal for a Legal Regime* (Martinus Nijhoff/Brill, 2009). He is Member of the International Institute of Space Law, and his main areas of research are space law, European union law and public international law.

F. Tronchetti, *Fundamentals of Space Law and Policy*,
SpringerBriefs in Space Development, DOI: 10.1007/978-1-4614-7870-6,

Made in the USA
Middletown, DE
28 August 2017